Springer Series in Synergetics

Synergetics, an interdisciplinary field of research, is concerned with the cooperation of individual parts of a system that produces macroscopic spatial, temporal or functional structures. It deals with deterministic as well as stochastic processes.

L.A. Blumenfeld

Problems of Biological Physics

With 38 Figures

Springer-Verlag Berlin Heidelberg New York 1981

Professor Lev Alexandrovitsch Blumenfeld

Institute of Chemical Physics,
Academy of Sciences, Vorobjevskoje Shosse, 2-b
Moscow 117 334, USSR

Volume Editor:

Professor Dr. Hermann Haken

Institut für Theoretische Physik der Universität Stuttgart,
Pfaffenwaldring 57/IV,
D-7000 Stuttgart 80, Fed. Rep. of Germany

Revised translation of the Russian edition:
Probleme biologitscheskoi phisiki (Problems of Biological Physics)
by L. A. Blumenfeld
© by Nauka, Moskow (1977)

ISBN-13: 978-3-642-67853-0   e-ISBN-13: 978-3-642-67851-6
DOI: 10.1007/978-3-642-67851-6

Library of Congress Cataloging in Publication Data. Blumenfeld, Lev Aleksandrovisch. Problems of biological physics. (Springer series in synergetics ; vol. 7) Translation of Probleme biologitcheskoi phisiki.
Includes bibliography and index. 1. Biophysics. I. Title. II. Series. QH505.B5513   574.19'1   80-24565

Softcover reprint of the hardcover 1st edition 1981

2153/3130-543210

Preface to the English Edition

This book was written, to a not inconsiderable degree, on the basis of the course "The Problems of Modern Biophysics" which the author gives to the students and postgraduates of the Biophysical Department at Moscow University School of Physics. It is meant for those who have a sufficiently good background in physics as well as in biology. I have tried to make this book intelligible to a broader circle of readers, i.e., to physicists not competent enough in biology, and to biologists not competent enough in physics. I hope that I have succeeded.

This book is neither a textbook nor a systematic account of a field of science. I think that in modern biological physics, i.e., in the branch of biology where people having fundamental physical or physico-chemical education are working, so many specific answers have been recently obtained that it is now just the right time to ask at least several questions of a general nature. The aim of this book is to formulate such questions though their choice is, to a considerable degree, determined by the authors preferences and interests.

Although, as it can be seen in the table of contents, this book deals with a set of biophysical problems, all of them essentially are different aspects of one big problem, the most interesting biophysical problem (at least from my point of view), viz. the relationship between statistics and mechanics in biological systems. In my opinion this problem is equally urgent both for modern physics and modern biology. Extensive studies of biological objects at the macromolecular level for the first time compelled physicists to consider systems which do not lend themselves to a separate statistical or mechanical description. As a matter of fast, this monograph is only an analysis of conclusions following from this statement.

The author's "Biological Weltanschauung" was developed as a result of many years of contacts and conversations with N.V. Timofeev-Resovskij. I wish to express my deep gratitude to Nikolay Vladimirovich for all I have learned from him.

I prepared the English translation myself. This was done not because I am too sure of my English but because I hope that I understand my Russian text better than most of the would-be translators.

Moscow, October, 1980   *L.A. Blumenfeld*

Contents

1. Introduction

1.1 What is Biophysics?

In this book some fundamental problems of biological physics will be discussed. It will probably be proper to begin with certain general remarks concerning the subject matter of this science.

The word "biophysics" is quite popular nowadays. Many journals in many countries are printed under this title, there exist research institutes of biophysics and departments of biophysics in universities, there are scientists calling themselves "biophysicists". This word, however, may lead and often does lead to a misunderstanding. Is there indeed "the biological physics", i.e., the special physics of living matter with laws differing from those of conventional physics as studied in higher schools and universities? The answer to this question is, to some extent, "a symbol of faith" and cannot be substantiated in its entirety. Nevertheless, every scientist who spends or intends to spend his time on the exploration of the living state must give some kind of an answer, at least to himself.

Therefore, I begin with the formulation of my answer, of my symbol of faith and shall try to substantiate it in this book, though, I repeat, it cannot be done unambiguously. *The known principal laws of physics are quite sufficient for a complete description and understanding of the structure and functioning of all existing biological systems.*[1]

Two points should be stressed. In the first place, we are speaking about the principal laws of physics. The situation here is the same as when we describe the structure and the functioning of complex mechanical or electronic devices. The law of behavior of these devices may be very specific and complicated, but nobody will try to develop new laws of physics in order to describe, for instance, the properties of a specific TV set circuit. These properties may be governed be quite definite and complicated rules, but they can be completely understood and derived on the basis of the physical principles governing the functioning of circuit elements and the knowledge of its construction and organization. In order to describe every new wireless set there is no need to create new physics.

---

[1]The laws of chemistry are not mentioned, because the foundation of these laws does not, in principle, require any new physical postulates.

Secondly, in this book biological systems will be considered as given a priori, as already existing. It means that in the realm of the symbol of faith formulated above, the science of physics is today unable to solve the problems of chemical and biological evolution. The theory of biological evolution, whose foundation was built by Darwin, is the major part of biology, of the science of living matter.

At the same time, formation of the existing biological systems as a result of chemical and biological evolution fully determines the physical and chemical mechanisms, structures and peculiarities of biological objects that will be discussed in this volume.

Thus, I think, biological physics can be defined as such a part of biology, where today one can work using the principles and methods of physics. In other words, biophysics is the biology for scientists with an educational background in physics. It is clear that biophysics is, so far, a minor and not the most important part of biology. According to the optimistic belief of the author, this part will grow continuously, asymptotically approaching the whole. Biophysics today is the part of biology which deals with the physical aspects of the structure and functioning of certain comparatively simple biological systems, without considering their origin and evolution.

We must now specify the meaning of the words "... complete understanding of the structure and functioning of all existing biological systems".

When a scientist states that he understands the chemical properties of benzene it means that it is possible for him, if only in principle, to cover logically all the distance from the postulates of quantum mechanics, electrodynamics, and statistical physics to the explanation of the reactivity and other chemical properties of the complex system containing 6 carbon nuclei, 6 protons and 42 electrons. Practically, nobody actually tries to cover the whole distance (and, due to immense mathematical difficulties, it cannot be covered without crude approximations), but there undoubtedly will not arise the necessity to introduce new physical postulates not deducible from those already existing. Certainly, when this path is being traversed, a great number of chemical laws are formulated, some of them of great generalization power, but every time one can cover the whole distance without new postulates.

It must be emphasized that it is not only senseless but even harmful to pass the whole way from the basic postulates to concrete explanations. It would mean a conscious refusal to use many rules formulated by chemists which, in the end, constitute the science of chemistry. The boring and, to a considerable degree, meaningless question about the possibility of reducing the laws of chemistry to those of physics can be, probably, answered in such a way: it can be done but it is rather unnecessary.

To be sure, one can encounter quite different situations in science. The understanding of electrical and magnetic phenomena on the basis of mechanics is im-

possible without the introduction of new postulates. These postulates are generaliz-
ations of the empirical experience and cannot be logically deduced from the laws
of mechanics. The introduction of new postulates seems to be always connected with
new world constants (in the above example — with the c constant). When we pass from
the laws of macrophysics to those of microphysics, we have to introduce the postulates of quantum mechanics and a new world constant, h.[2]

The symbol of faith mentioned above is essentially the conviction that in biophysics, as in chemistry, it will not be necessary to introduce new postulates and
world constants in order to understand the structure and functioning of biological
systems.

At the same time, as already stated, the basic chemical and physical processes
in biological systems do not on any account exhaust the whole of biology. In the
foundation of biology lies the concept of biological evolution. It does not mean,
of course, that in order to understand biological evolution we shall be obliged to
introduce new physical postulates. The primary event of evolution is the convariant
reduplication, i.e., the self-reproduction of biopolymer complexes carrying the
genetic information, with the reproduction of structural variants, occurring by
means of chance mutations [1.3].

The future biophysical theory of biological evolution must take into account
the essential difference between the phenotype and the genotype, i.e., between the
organism interacting with its environment and the instruction for its self-reproduction which is hardly at all subject to the influence of external conditions.

Mutations, i.e., the elementary changes of heredity, take place only in the
genotype, while environmental alterations may lead only to nonhereditary changes
in the phenotype. The connection between the phenotypical and genotypical changes,
i.e., the successive changes in the genotype determined by the environment, are
realized by means of the Darwinian mechanism of natural selection. As a matter of
fact, it is the genotype that undergoes evolution, but it becomes apparent as the
changes in the phenotype.

In trying to construct the theory of biological evolution one can, of course,
in principle put aside the question of how the mechanism of instruction written
in the genotype is realized, and how the genotypical changes are transformed into
the phenotypical ones, i.e., the question of the ontogenesis mechanism. The current mathematical theories of biological evolution are doing just that. In view of
the low level of present-day knowledge, it is probably quite justified. However,
the separation of evolution and ontogenesis reduces the evolution problem to
certain mathematical constructions and leaves some feeling of dissatisfaction.

---

[2]For the two types of physical theories, those requiring and those not requiring
new postulates and world constants, see also [1.2].

Some people, full of enthusiasm as regards the recent achievements of molecular biology, think that the development of the biochemistry of nucleic acids and nucleoproteides, the clarification of phage DNA replication, of protein synthesis and its regulation, are leading us steadfastly to the true understanding of the mechanisms of ontogenesis and elementary hereditary phenotype changes. Here, however, one should bear in mind that in this branch of biology all we know at the molecular level concerns only the mechanisms of hereditary changes in protein molecules. It is a long way from proteins to characters, and this way has not even been explored as yet. At the same time, it is necessary to traverse this way in order to understand evolution. I think that the solution of this, probably the most important problem of modern science, will require, first of all, new ideas. It is rather a pity that these ideas will probably belong to that part of biology which is not yet considered biophysics.

Attempts at a thermodynamical approach to biological evolution are discussed in Chap.2.

There is one essential difference between the situations in biophysics and chemistry. A chemist, as a rule, knows the logical "way" from the structure of his object to its chemical properties, and the difficulties arising are usually of technical origin. A biophysicist, in most cases, does not know the way from the structure to the function. He is only sure that such a way does exist.

1.2 Cell Components and Their Specific Features

Hitherto we have been speaking about "biological systems", or "living matter" without defining these objects more precisely. In biology the distinction between the living and nonliving is rather arbitrary. Most biologists think that life begins from the cell. In our discussion of biological systems in this book we shall but seldom overstep cell limits. We, therefore, accept this arbitrary criterion, and when not stated otherwise, "biological system" or "living matter" means just the cell.

First of all, it must be stressed that a cell contains the same chemical elements as does any object of dead nature. If we consider the elementary low-molecular units contained in every cell, — aminoacids, nucleotides, coenzymes, etc. — it becomes clear, that there are no principally new compounds, no building blocks, used only in the living systems. Identical or very much similar compounds are synthesized and studied by chemists dealing with essentially nonliving objects.

A more detailed analysis of these low-molecular "bricks" of living matter gives evidence of several interesting general characteristics. Although these characteristics are not exclusively specific to living matter, their existence can lead to meditations on the possibility for these properties to play an important role in the functioning of living systems.

The principal low-molecular cell components — nucleotides, porphyrins, flavines, quinones, some aminoacids, carotenoids, etc. — have several common properties. They are characterized by comparatively low[3] electron excitation energy, low ionization potential, high electron affinity, high electron polarizability. To make a long story short, it seems that all these chemically active groups are destined for taking part in electron transfer processes.

There is another interesting peculiarity of the elementary low-molecular building blocks of the cell. The same chemical structures take part in quite different processes. Pyridine nucleotides (NAD and NADP), working within the cell as hydrogen carriers, contain nitrogen bases, residues of ribose — a five-membered sugar, and phosphoric acid residues. Adenosine triphosphate (ATP), that performs within the cell the function of energy transformation and storage, contains also a nitrogen base, phosphoric acid and ribose residues. The monomers in the nucleic acids, intended for hereditary information storage and transfer and performing protein synthesis, are built of the same compounds. Such "lack of imagination" in nature is rather surprising. If a chemist had to synthesize a special compound for hydrogen transfer, he, probably, would do it quite differently. One has the impression that nature for different purposes has just used the materials at hand. According to this point of view, there is no particular uniqueness in the concrete chemistry of cell low-molecular compounds.

It could be, in principle, possible to build living matter of different low-molecular bricks. The uniqueness of today's chemistry of living matter is due to the fact that it has already been realized and fixed by billions of years of chemical and biological evolution.

One can often hear that the most important thing is not the specific chemical composition, but the unique sequence of monomers in the protein and nucleic acid chain molecules. There is some truth in this statement. We know, for instance, that the nucleotide sequence in all existing DNA molecules is fixed. This sequence determines the structure of all cell proteins, and, practically, most of the properties of unicellular, and, probably, of multicellular organisms. Can we, however, conclude, that only these existing unique sequences are compatible with life?

Let us perform a rather primitive calculation and find out the probability of spontaneous formation of at least one of the presently on earth existing monomer sequences in the DNA molecules. As a result of this calculation, we obtain the upper limit of this probability because all the approximations that will be used can only increase the probability value.

Let us assume at first that all the presently existing nucleotide sequences are unique, not in a sense that only these sequences were fixed in the course of chemical and biological evolution, but that they are chemically unique, i.e., that

---

[3]Low and high — in comparison with the majority of organic compounds.

only these sequences ensure such properties of biological structures which are compatible with life. What is the number of such sequences? Let us assume that the layer of the biosphere on the earth's surface is 1000 m thick and packed with DNA molecules having the molecular weight of 10^6 and density of 1. The number of DNA molecules in the biosphere will then be $\sim 10^{41}$. Let us further assume that all these DNA molecules have different nucleotide sequences, i.e., that there exist 10^{41} unique sequences compatible with life. The number of all possible sequences in a two-stranded DNA molecule with MW 10^6 (the nucleotide number in one strand is ~ 1500) and containing four different types of nucleotides, equals $4^{1500} \approx 10^{900}$. Assume now that during the whole time interval of the earth's existence ($\sim 10^{10}$ years or $\sim 10^{17}$ s) on each square nanometer of its surface (this surface is $\sim 10^{32}$ nm^2), every 10^{-8} s one "attempt" was accomplished, i.e., one polynucleotide molecule having one of the possible sequences way synthesized. The probability that during this time at least one DNA molecule with "unique" sequence (one out of 10^{41} existing) will be formed is equal to $3 \times 10^{17} \times 10^8 \times 10^{32} \times 10^{41} \times 10^{-900} \approx 10^{-800}$.

It may appear that this calculation is founded on the assumption of simultaneous polymerization of 1500 nucleotides. This is not so. The same result would be obtained if the synthesis were carried out according to an arbitrary mechanism with a gradual increase of molecular weight (naturally the time interval per each "attempt" would be in this case greater, and the resulting probability smaller). The only requirement is the absence of any essential correlation between the nucleotide sequences along the chain. According to our present knowledge about nucleotide sequences in DNA and amino acid sequences in proteins this requirement is fulfilled.

The calculation result implies two contradictory conclusions: 1) The rise of life is impossible. 2) The postulate about the chemical uniqueness of existing sequences is erroneous.

The first conclusion is evidently wrong.

Consequently, we must conclude that monomer sequences in biopolymers realized in living nature differ from other possible sequences only by the fact of their existence, and their chemical and physical properties are not unique in themselves.

The uniqueness of biopolymer structure is thus, essentially, biological. It has become the only one possible due to the long biological evolution. Apparently, the chemical structure of low-molecular cell components and biopolymers should satisfy only a comparatively small number of general requirements, and the result (i.e., the rise of living matter) has been but slightly dependent on the concrete chemical structures and physical characteristics, realized at random during the early stages of evolution. There could be, in principle, other building blocks and other sequences!

1.3 The Aim of the Book

As already mentioned, this book deals with the basic problems of biological physics. Biological systems will be considered as a priori given, having come into
being due to the evolutionary process. We shall discuss here not only the problems
that have been more or less solved, or have not been solved at all (most of them
haven't), but also the pseudoproblems, the questions that are meaningless, despite
numerous pages devoted to them in current literature.

The following three chapters deal with thermodynamical aspects in biology on
which there exist many contradictory views. The analysis of experimental data and
theoretical constructions leads to the conclusion that biological systems and
their components (beginning from the macromolecular level) are mechanical-statis-
tical machines. The mechanics ensures sufficiently large relaxation times necessary for all important intracellular processes, and statistics ensures a high degree of reliability.

In the remaining chapters the physical aspects of conformational and configurational changes in biopolymers, enzymatic catalysis, electron transfer and energy
transformation are considered. The author believes that these phenomena determine
the majority of the processes within a living cell. The choice of particular
examples is determined exclusively by the author's knowledge and personal interests.
The fact that protein biosynthesis and molecular genetics are not even touched
upon here certainly deserves dissaproval. The only justifaction can be found in the
existence of an enormous quantity of reviews, popular papers and books on these
subjects, that already approaches the saturation level.

2. The Ordering of Biological Structures

2.1 Are They so Ordered in Reality?

Almost in every book on the theoretical aspects of biology and biophysics one can find statements about the surprising ordering of biological structures at all levels of organization — from macromolecule to organism. In his brilliant book "What is Life?" [2.1] that was so stimulating in the creation of modern biophysics, SCHRÖDINGER stated that living objects "feed on negative entropy". Meditations on the "antientropic tendencies of life" and even on the "antientropic character of biological evolution" are now the common trend not only among biologists and philosophers but among physicists as well.

Indeed, when a scientist observes certain biological structures and processes (cell division, morphogenesis) he is usually surprised by their spatial and temporal ordering. Surprise, however, is an emotional category. To be able to state that one system is more ordered than another, one must measure the corresponding degrees of ordering. There is only one measure of the degree of ordering in an object: its entropy.

In this chapter a comparison of entropy values for living and nonliving systems will be made. It must give us the estimation of entropy changes associated with the formation of biological organization.

These comparisons will show that there does not exist any special high degree of ordering in biological systems. Living matter does not possess any "antientropic tendency". Therefore, it will be necessary to clarify what is the objective sense of vague ideas about the special ordering of living systems. In this connection a very delicate question of the meaning of biological ordering will be raised and discussed.

2.2 Entropy and Information

"I have thrown a die five times and each time have had two". "From a pack of 32 cards I go one card at random, and it was the ace of hearts". "It rained in Moscow yesterday". Which of these statements contains the most information?

The problem of estimating the absolute information content in a communication was solved in 1949 [2.2].

The information content is determined as[1]

$$I = \lg_2(P_1/P) \tag{2.1}$$

where P denotes a priori probability of a certain event (the probability before the communication), and P_1 the probability of the same event after the communication. For the sake of simplicity we shall assume that all communications are trustworthy, i.e., $P_1 = 1$. Then

$$I = \lg_2 P \quad . \tag{2.2}$$

According to (2.2), the information content of a trustworthy communication about an event with a priori probability of 1/2 equals 1. This unit of information content is called "bit" (binary digit).

In the first statement $P = (1/6)^5 = 1/7776$; $I = \lg_2 7776 \simeq 12.9$ bits. In the second statement $I = -\lg_2(1/32) = 5$bits. For the third estimation we need the a priori probability of rainy weather in Moscow. According to statistics, the average number of rainy days in Moscow is 170 per year. Thus, $I = -\lg_2 170/365 \approx 1.1$bits.

Twenty years before SHANNON the information problem was analyzed by Szillard in connection with the so-called MAXWELL paradox [2.3]. The essence of this paradox can be formulated as follows. An isolated system includes a gas-filled box divided into two parts by means of a partition with a door. A "demon" is placed within the box near the door. This demon, a living being or an automat, is capable of seeing individual gas molecules and telling fast molecules from slow ones. The demon opens the door only to fast molecules coming from the right. Therefore, the temperature of the gas in the left part of the box will be rising and that in the right one falling. In an isolated system the heat is thus transferred from the cold body to the hot one, and the entropy decreases in contradiction with the second law of thermodynamics.

Several generations of physicists have tried to solve this paradox. However, it was SZILLARD [2.4] who, discussing one of the simplified variants of the demon paradox (vessel with a piston contains only one gas molecule), called special attention to the necessity of getting information about the molecules and, thus, found the connection between information and thermodynamics. After SZILLARD, great contributions to our present-day understanding of this connection have been made by DEMERS [2.5,6], ROTHSTEIN [2.7,8], and BRILLOUIN [2.9].

Let us now consider the solution of the Maxwell paradox according to Brillouin. The demon can see the molecules only if he (or the part of the device called demon)

---

[1]This formula is valid for a simplified situation, when 1/P events of equal probability can take place.

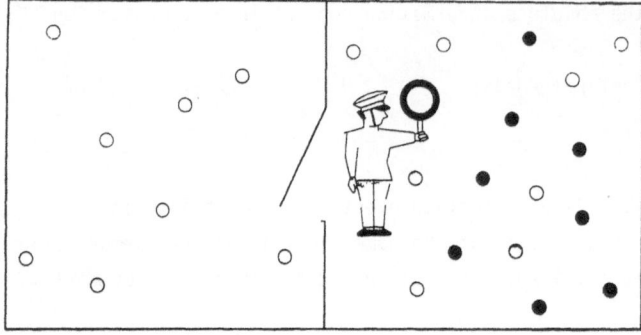

Fig.2.1. Maxwell's demon

is out of thermodynamic equilibrium with the rest of the system. Otherwise the
demon would be able to see only the intensity of black-body radiation, whose spectrum depends only on the temperature. The demon, therefore, must be provided with
an electric torch, and the filament temperature must differ from that of the system
(the demon has to illuminate the molecules!). A detailed analysis shows that oper-
ations connected with information transfer lead to the compulsory increase in system entropy by at least (in the ideal case) the value equal to the entropy decrease
due to the demon's activity.

Many authors discussing these problems (especially in biology) use the term negentropy, the meaning of which is just the entropy with the negative sign. It is difficult to understand the purpose of it. Why should it be preferable to speak about
the negentropy increase instead of the entropy decrease? After all, negentropy is
essentially the Boltzmann H function. However, for historical reasons this function
is usually taken with the opposite sign and called entropy. It is rather difficult
to see a reasonable substantiation of the new sign change and the introduction of
a new word.

The relation between entropy and information was formulated by Brillouin in the
form of the so-called *negentropy principle of information.*

a) Information can be transformed into the entropy decrease of a system and
vice versa (i.e., information can be used to decrease the system entropy, and the
entropy decrease, to obtain information). In the ideal case of reversible processes, these transformations may be, in principle, carried out without losses.

b) In the course of any experiment that increases our knowledge of a system
(i.e., that gives us additional information about this system), the entropy of
the system, or that of the environment, increases. This increase is equal to (for
reversible processes) or is greater than (for irreversible processes) the entropy
decrease of the system due to the information obtained.

The relation between entropy and information may be understood in the following
way. Let us fix the macroscopic state of some system, i.e., let us fix with a definite degree of precision such parameters as volume, pressure, temperature, chemi-

cal composition, external magnetic, electric, and gravitational fields, etc. For each macroscopic state of a system there exists a multitude of microscopic states. In a microscopic state the coordinates and momenta of all the particles within the system are defined with precision allowed by the uncertainty principle. Let the number of microscopic states, corresponding to the given macroscopic state, be W (for any macroscopic system at a temperature higher than OK this number is enormous). W is the thermodynamic probability or statistical weight. All W microscopic states, corresponding to the given macroscopic state, have an equal a priori probability. To know what is the microscopic state of a system, is to know all about this system. Let us try to calculate, how much information needs to be obtained in order to define the microscopic state of a system in the given macroscopic state. In other words, how much information is lacking for a total description of a system in the given macroscopic state?

Let us assume that we have carried out a series of experiments and defined the microscopic state[2]. Before this series of experiments the a priori probability for the system to be in the given microscopic state had been 1/W, and after the experiments were completed the probability became 1. The information obtained is, according to (2.2),

$$I = -lg_2(1/W) = lg_2 W \ .$$ (2.3)

At the same time, according to the Planck-Boltzmann formula, the entropy of the system, the thermodynamic probability of which is W, equals

$$S = k \ln W$$ (2.4)

where $k = 1.38 \times 10^{-16}$ erg $\times$ deg^{-1} = 3.31×10^{-24} entropy units (1 e.u. = 1 cal $\times$ deg^{-1}) — the Boltzmann constant.

It is easy to see that (2.3) coincides with (2.4) with an accuracy determined by the constant dimensional factor. Thus, the system entropy is essentially the information lacking for its complete description. I and S are one and the same quantity. The situation here is just the same as in the case of the total energy of a particle and its mass, interrelated by formula $E = cm^2$ with a constant factor c^2, or as in the case of the energy of a quantum and the corresponding radiation frequency ($E = h\nu$, h being the constant factor). In order to convert the information in bits into the entropy in entropy units, it is necessary to convert the logarithm to the base 2 into the natural logarithm and to multiply it by k. Thus,

$$S[e.u.] \cong 2.3 \times 10^{-24} \ I[bits] \ .$$ (2.5)

Some scientists are afraid of such reasoning. They think that acknowledging the equivalence of the system entropy and the information lacking for its complete description may lead us to the acknowledgement of the dependence of an objective

---

[2]It cannot be done in actual fact.

physical property — the system entropy — on a subjective factor, the extent of our knowledge. These fears are unfounded. The degree of uncertainty of a microscopic state in fixed macroscopic state of a system is quite an objective characteristic. It increases, for instance, with temperature, due to the increase in the number of possible microscopic states. If an explorer tries to lower the entropy by means of experiments leading to a decrease in the number of possible microscopic states (measuring, for example, the macroscopic parameter with greater accuracy), the perturbations due to the measurement process must then, according to the Brillouin principle, lead to the entropy increase, whose value will be at least equal to the entropy decrease due to the additional information about the system.

It is of interest to note that, for a macroscopic system, any imaginable in-crease in the accuracy of a macroscopic parameter measurement leads to a negligible entropy decrease. Let us consider a simple example. The entropy change due to the evaporation of one gram of water is ~ 1.8 e.u. at $40^{\circ}C$. Let us make the macroscopic state more definite and increase the temperature measurement accuracy from $\pm 10^{\circ}C$ to $\pm 10^{-6 \circ}C$. It is quite a good approximation to assume that the number of possible microscopic states is proportional to the temperature interval. Therefore, as a result of the increased precision of temperature measurement, the number of pos-sible microscopic states decreases by a factor of $20/2 \times 10^{-6} = 10^{7}$. The information lacking for the complete description of our system decreases by $lg_2 \, 10^7 \approx 23.3$ bits, and the entropy by $\sim 5 \times 10^{-23}$ e.u., i.e., by a negligible fraction of the measured value!

Another remark. Strictly speaking, the concept of entropy and thermodynamic probability have a meaning only for equilibrium states. Therefore, in the case of biological systems, the numerical values of entropy must be considered in the same way as in nonequilibrium thermodynamics (see Chap.3).

2.3 What is the Cost of Biological Ordering?

Many chemical processes occurring within the cell are accompanied by essential entropy changes (increases as well as decreases). These changes do not differ in principle from the corresponding changes taking place in the course of conventional chemical reactions. They have no biological specificity and are not of interest to us here. In this paragraph we shall calculate the entropy changes due to essen-tially "biological" ordering, i.e.: 1) the ordering of the formation of a multi-cellular organism from individual cells; 2) the ordering of the formation of a cell from biopolymers; 3) the ordering of the formation of proteins and nucleic acids (we shall restrict ourself to DNA) from corresponding monomers. These cal-culations will, of course, be extremely approximate, but they may answer the fun-damental question: are there any thermodynamic criteria of biological ordering?

1) *The ordering of the formation of multicellular organisms from cells.* The human body has about 10^{13} cells. Let us assume that there is no pair of identical cells and that no pair of cells is interchangeable. It means that relative cell positions are unique. The information necessary for the building of only one structure out of $(10^{13}!)$ possible ones, is

$$I = lg_2(10^{13}!) \simeq 10^{13} \, lg_2 \, 10^{13} \simeq 4 \times 10^{14} \text{ bits} \quad .$$

The entropy decrease of human body formation due to this type of ordering is thus

$$\Delta S \simeq 2 \times 10^{-24} \, 4 \times 10^{14} \simeq 10^{-9} \text{ e.u.}$$

2) *The ordering of the formation of a cell from biopolymers.* The number of bio-polymer molecules (proteins, nucleic acids, phospholipids, etc.) within the cell averages 10^8. Let us again postulate that all molecules are different, and their relative positions are unique. The information necessary for the building of one cell from ready-made polymers is then

$$I = lg_2(10^8!) \simeq 10^8 \, lg_2 \, 10^8 \simeq 2.6 \, 10^9 \text{ bits} \quad ,$$

and for all the cells in the human body $\sim 2.6 \times 10^{22}$ bits, corresponding to the entropy decrease $\Delta S \simeq 6 \times 10^{-2}$ e.u. (this value also accounts for the previously calculated ΔS of organism formation from cells).

3) *The ordering of the formation of proteins and DNA from monomers.* The human body contains about 7 kg of proteins and about 150 g of DNA, or $\sim 3 \times 10^{25}$ aminoacid and $\sim 3 \times 10^{23}$ nucleotide residues. In order to build a unique sequence among $20^3 \times 10^{25}$ possible for proteins, it is necessary to obtain 1.3×10^{26} bits of in-formation. In the case of DNA this number equals 6×10^{23} bits. In entropy units, these values are ~ 300 and ~ 1.6 (the two types of ordering treated earlier are automatically accounted for in these values).

Thus, the ordering of the biological organization of the human body is not higher than 301.5 e.u., the main contribution being made by the ordered distri-bution of aminoacid residues in protein molecules. This entropy decrease due to the origination of one of the most complicated biological organizations — the human body — can be compensated by trivial physical and chemical processes. For instance, the entropy increase by 300 e.u. is provided for by the evaporation of 170 ml of water or by the oxidation of 900 g of glucose.

These estimates show that formation and development of biological organizations proceed thermodynamically almost free of charge. The entropy of 10^{13} different unicellular organisms is nearly the same as that of the human being containing 10^{13} cells.

All talk about "the antientropic tendency" of biological evolution, about the unique ordering of living matter is based on a misunderstanding. According to thermodynamic criteria, the ordering of any biological system is not greater than that of a piece of rock having the same weight.

2.4 The Meaning of Biological Ordering

Although the reasoning and calculations of the preceding paragraphs are difficult to disprove, there remains the feeling of a certain dissatisfaction. It is hard not to think that there is some deception: this remarkable ordering of biological processes is quite evident without any calculations! If physics states that this ordering is negligible, one wants to say: so much the worse for physics! It is intuitively clear that biological structures maintain their specific, qualitatively different ordering, which can be measured, of course, in terms of entropy units, but such estimations cannot give us anything valuable for the understanding of the specificity of living matter.

It is easy to see that this intuitive feeling of there being a specific ordering of biological structures is based on the fact, that this ordering is meaningful. The ordering of biological matter, the information contained in living objects, has a meaning. A meaning is a teleological conception: to have a meaning is to have a purpose. The muscle, nervous and other cells of the heart are distributed in a specific order for the purpose of the proper functioning of this organ. The molecules of various proteins and phospholipids are positioned in an orderly way within the mitochondrial membranes for the purpose of ensuring a high efficiency of the electron transfer and energy accumulation processes in these membranes. The aminoacid residues in a γ-globulin molecule are positioned in an orderly way for the purpose of ensuring the specific immunological activity of this protein.

Let us try to elucidate the meaning of biological ordering, to clear up what must be the characteristics of systems the ordering of which has a meaning.

I think that these questions were for the first time raised by QUASTLER in his short and rather remarkable book: "The Rise of Biological Organization" [2.10]. We use here the principles of QUASTLER's approach to this problem and some of his examples.

In a certain large water reservoir nucleotides, the building blocks of desoxyribonucleic acid (DNA), are dissolved. There are four types of nucleotides in solution: desoxyadenosine-5'-phosphate (A), desoxyguanosine-5'-phosphate (G), desoxycitidine-5'-phosphate (C), and desoxytymidine-5'-phosphate (T). The nucleotide molecules are involved in polycondensation reactions leading to the formation of di-, three-, and other polynucleotides (unithread polymers):

1.   HO—P—O—... A   +   HO—P—O—... T   →

→ H_2O + HO—P—O—... (structure with R_1 and R_2)

Here R_1 and R_2 are adenine and thymine residues, respectively.

2.   AT + A H_2O + ATA,   etc.

Here R_1 and R_2 are adenine and thymine residues, respectively.

The hydrolysis rate of phosphodiether bonds being formed during polycondensation is much higher than their formation rate. The equilibrium state of this system must, therefore, be shifted toward monomers. The momentary concentrations of di-, three-, and other polymers must be very small and become less with polymer length.

Let the rate constants of the reactions

$$(N)_i + N \rightarrow (N)_{i+1} + H_2O ,$$

where N denotes any one of the four nucleotides, be independent of the chain length and of the type of nucleotides to be condensed. Moreover, let the concentrations of all nucleotides (A,G,C,T) be equal. Then, small amounts of polymer chains of varying length, which will always be present in dynamic equilibrium in the solution, must have random sequences, and the a priori probability of all sequences will be the same. If the conditions of rate constants and concentrations being equal do not hold true, some sequences will occur more often and there will exist a certain distribution of a priori probabilities of the sequences. However, even in this case, any sequence arising due to the reversible polycondensation will be meaningless.

Consider now a more complicated system. In addition to polycondensation and hydrolysis processes (i.e., the reactions of synthesis and disruption of unithread polymers), there, in principle, can proceed the matrix synthesis by means of complementary nucleotide sorption on a polymer thread, and phosphodiester bond formation between nucleotides already adsorbed on the polymer chain. Nucleotide sequence in the new chain is completely determined by the nucleotide sequence in the original unistrand polymer: opposite A is always T, and opposite G,C. It is determined by the stereochemistry of corresponding nucleotides. As a result of such matrix synthesis

$$(N)_i + iN \rightarrow [(N)_i]_2 + i\ H_2O$$

a two-strand polymer arises, which is much more stable, than the unistrand one. It practically does not undergo any hydrolysis. The hydrolysis

$$[(N)_i]_2 + 2i\ H_{2O} \rightarrow 2iN$$

is much slower, than the dissociation (splitting) and than the matrix reduplication

$$[(N)_i] + 2iN \rightarrow 2[(N)_i]_2 + 2i\ H_2O\ .$$

The first two-strand molecule in this system arises as a result of a random process of very low probability: on one of the unistrand polymer molecules the matrix synthesis had time to be completed before the polymer hydrolysis. The nucleotide sequence in this unistrand molecule may be completely arbitrary. After the formation of a two-strand structure, however, the situation in the system changes abruptly. The nucleotide sequence existing in such long-living two-strand polymers is now meaningful. This meaning is quite obvious: in a stable two-strand molecule capable of reduplication this sequence exists, but other, in principle allowed sequences, do not. Due to specific properties of the two-strand structure, mentioned above, the concentration of unistrand polymer molecules, having this, now specific, sequence, will increase. These unistrand polynucleotides are now in dynamic equilibrium not only with monomers, but also with stable two-strand structures having the same nucleotide sequence. The two-strand polymer concentration will increase on account of monomers due to the reduplication process.

The accidentally arisen unistrand polymers with the "proper", "meaningful" sequence have now big chances to remain "living", forming stable structures with complementary polymer molecules, but unistrand polymers with "bad", "meaningless" sequences will certainly undergo hydrolysis. The small chance deviations from the "right" sequence, which do not diminish appreciably the two-strand structure stability, will also be reduplicated and give rise to the separate systems competing with the first one for the monomer nucleotide stock available.

Thus, owing to *the memorizing of the random choice* in the hypothetical example given above, a meaningful ordering has arisen, a system capable of creating meaningful information has been made. We can now ask: what for do the polymers in this system have this nucleotide sequence?; and get the answer: in order to exist, because with this sequence they have the maximal probability to remain "living". The physical properties of polymer molecules having this sequence may not differ from those of other polymers. This sequence has a meaning just because a self-reproducing polymer system with this sequence has accidentally arisen. The meaning of ordering is, thus, a biological category, i.e., it is determined by all the history of the formation of a system, by its evolution. It does not mean that the concept of "meaning" is somewhat mystical and does not obey the laws of physics. All properties of meaningful objects can be logically derived from their physical characteristics, provided we know their history, their evolution.

As it was already mentioned in Chap.1, the author thinks that the problem of evolution is not, so far, a part of biological physics.

The above example of a stable self-reproducing nucleic acid system is not claimed to be connected with the origin of life on our planet. It is just an attempt to demonstrate the most essential characteristics of the meaningful ordering creation process using well-known biopolymers as an example.

QUASTLER describes another example of the rise of meaningful ordering. Imagine a safe with a lock that can be opened with the help of a three-figure code. There is a set of figures from 0 to 9 and a device for random choice allowing to select three figures from the set. Let us choose three figures and introduce them as a code into the safe lock. Before this operation, any imaginable sequence of three figures did not differ from any other: all sequences were meaningless. Introducing three figures chosen at random into the safe lock we have given a meaning to this sequence. This meaning is very simple: this sequence can open the safe, others cannot.

It is easy to see that this example does not differ in principle from the example with nucleic acids. In both cases, the meaningful ordering, the new meaningful information was created by means of memorizing the random choice. We must speak only about the creation of new information, but not about its extraction from the noise. Prior to the formation of a stable two-strand structure by a polymer with random nucleotide sequence, or prior to the introduction of a random three-figure sequence into the safe lock, there was no information about which of these sequences is preferential (one is "better" because it had acquired the ability of prolonged existence and self-reproduction; the other is "better" because it had acquired the ability to open the safe). The ordering has obtained a meaning, information has been created. As a result of this procedure, the unforeseeable has turned into the inevitable. Composer Pierre Boulet said that the transformation of the unforeseeable into the inevitable is creation [2.10]. It is possible that the principle of memorizing the random choice is the foundation of any type of creative activity.

Ability to create new information, to give a meaning to the ordering, is, probably, a necessary property of living matter. Any definition is limited and conditional, but I cannot refrain from the following formula. We call those systems "living" which are self-reproducing and capable of creating information that either directly or indirectly influences their self-reproduction.

2.5 The Necessity of Mechanical Details

Systems able to create a meaningful ordering, the examples of which were given in the previous section, have one common property. They contain meaningful components, constructions whose lifetimes are greater than the working cycle times of these systems. In the case of a nucleic acid system it means that the stable component — the two-strand polymer — does not dissociate and undergo hydrolysis prior to the reduplication process, and in the second example, that the lock does not disintegrate after code introduction and before at least one three-figure set is tested.

This requirement of long-living slowly relaxing structures (constructions) appears to be compulsory for living matter. It is impossible to build a biological system using only the gaseous phase!

Throughout this volume, when discussing the physical aspects of quite different biological processes, we shall always see, that slowly relaxing mechanical components — constructions — play a most important part in these processes.

2.6 The Problems

Summing up the content of this chapter, we can say that the problems concerning the ordering of biological structures, the origination and evolution of these orderings, do not belong to the realm of physics[3]. There are no thermodynamic aspects of biological evolution, and there exists no specific "biological thermodynamics".

It has recently become fashionable, when treating the questions of theoretical biology, to use the approach of the information theory. Up to now, however, most of the works using information theory in biology are rather trivial: well-known facts are translated into new language. For biological systems with a meaningful ordering it is the quality and not the quantity of information that is most important. The construction of the formal apparatus of information theory, taking into account the quality of information, would be a major step in the development of theoretical biology. I am not quite sure it can be done.

---

[3]If, of course, one does not believe that all phenomena in nature belong to the realm of physics.

3. Nonequilibrium Thermodynamics and Biological Physics

3.1 Open Systems

Biological objects are thermodynamically open systems able to exchange energy and
matter with environment. During the last thirty years this trivial statement has
been repeated so often that a susceptible reader or student begins to ascribe to
this statement a deeper (although rather vague) meaning that the one it has in
reality.

Conventional thermodynamics is strictly applicable only to isolated systems
containing an immense number of particles in the state of thermodynamic equilibrium.
It is only for such systems that the basic thermodynamic quantities — temperature
and entropy — have a rigorous physical meaning. It was already clear to Boltzmann,
and was unambiguously proved by CHINCHIN in his brilliant, but, regrettably almost
unknown, monograph [3.1]. Conventional thermodynamics is, therefore, just thermo-
statics.

No completely isolated systems exist in nature. Therefore, one must clearly
understand what meaning thermodynamic quantities have in real systems. We can use
the concepts of equilibrium thermodynamics in the cases when the time interval is
sufficiently long to allow the interactions between the system components to get
the system into the neighborhood of local equilibrium, and sufficiently short to
neglect the "slow" interactions between the system and environment (see [3.2]),
and in the case of slow changes in the degree of "kinetic nonequilibrium" (see
Chap.4).

In all cells and organisms, many processes take place which satisfy these requirements. They are, mainly, chemical transformations, provided we are interested
only in the initial and the final state in a process that was chosen with reason
among the multitude of interconnected and simultaneously occurring reactions. In
this sense, the application of equilibrium thermodynamics to living systems is as
regular (or as irregular), as to dead systems.

In the living and dead nature situations are, however, encountered for which
the interactions with environment cannot be reasonably neglected. To describe these
phenomena there exists a science called "thermodynamics of irreversible processes",
or "nonequilibrium thermodynamics". Strictly speaking, this theory is only appli-

cable to the "not too nonequilibrium" systems. This science is very useful for the treatment of many effects in the living and dead nature, and biological systems do not show any specificity in this respect.

The first attempts to develop nonequilibrium thermodynamics were made in the middle of the last century and were associated with certain thermoelectric phenomena. The basic equations of the theory, however, were obtained as recently as in 1931 by ONSAGER [3.3,4]. Further development of nonequilibrium thermodynamics is connected with the names of Casimir, Meixner, and Prigogine.

Apparently, the first ideas on the principal nonequilibrium state of living matter were set forth by BAUER [3.5] and GOORWITCH [3.6]. "Systems in a state of stable nonequilibrium" (Bauer) and "molecular constellations" (Goorwitch) are very close to the idea of stationary states maintained out of equilibrium by constant external forces (Onsager's nonequilibrium thermodynamics) and even to the idea of dissipative structures in Prigogine's works.

In this chapter, the principles of nonequilibrium thermodynamics in the region where Onsager's reciprocal relations hold true will be outlined (see, e.g. [3.7-9]). At the end of the chapter certain results recently obtained by Prigogine and his co-workers on dissipative structures far from equilibrium will be presented. In connection with these topics, it will be necessary to touch upon the oscillatory regimes in chemical and biological processes. Consideration of relevant works allows us to restrain certain qualms concerning the validity of the general thermodynamic approach to the behavior of systems far from equilibrium.

3.2 Phenomenological Relations

For any irreversible process in a macroscopic system two types of characteristic parameters can be specified. In the heat conduction phenomena, for instance, such parameters are thermal flux, Q, and temperature gradient, grad T, causing it. In the case of electric conductivity, flux of electricity (current, I) is proportional to potential gradient, grad V (Ohm's law). Fick's law relates diffusion flux to concentration gradient. In the case of chemical reaction, the flux of transformed compound (i.e., the reaction rate, v) is determined by chemical affinity $A = \sum_i \nu_i \mu_i$, where ν_i and μ_i are the stoichiometric coefficient and chemical potential for molecular species i, respectively, and the products $\nu_i \mu_i$ for initial and final reaction components enter the sum with different signs.

Concentration, temperature and potential gradients, chemical affinity, etc., causing matter, heat or charge transfer, chemical transformation, etc., in the system have been called thermodynamic forces, or just forces, and are as a rule, designated by symbols X and Y. The quantities of the second kind have been called thermodynamic fluxes, or just fluxes, and are usually designated by J. We have seen that the laws of the above-mentioned irreversible processes imply a proportionality between the fluxes and the corresponding forces

$$J_i = L_{ii}X_i \tag{3.1}$$

where L_{ii} denote the phenomenological coefficients (of heat conductivity, electric conductivity, diffusion, etc.).

As a general rule, any force is able to cause any flux. For instance, grad T may lead to the appearance not only of the heat flux, but of the flux of matter as well (thermodiffusion). Therefore, the linear phenomenological relations are to be written in the form

$$J_i = \sum_{k=1}^{n} L_{ik}X_k \; . \tag{3.2}$$

Let us consider now Onsager's theorem of reciprocal relations that form the basis of nonequilibrium thermodynamics. Our discussion will be restricted to the analysis of the physical meaning of the theorem and of the assumptions necessary for its proof. In order to formulate this theorem with greater precision, it is necessary to take into account the existence of two kinds of thermodynamic forces, designated here as X and Y. For the sake of simplicity, we exclude from our consideration the systems in the external magnetic field.

Consider a system, the state of which (and, consequently, the existing thermodynamic forces) is determined by parameters of two kind: A_1, A_2, ..., A_n — the even functions of the system particle velocities (for instance, local energies, concentrations), and B_1, B_2, ..., B_m — the corresponding odd functions (for instance, local velocity gradients). Let the deviations of these parameters from their equilibrium values A_i^0 and B_i^0, be

$$\alpha_i = A_i - A_i^0 \quad , \quad \beta_i = B_i - B_i^0 \; . \tag{3.3}$$

The entropy deviation (ΔS) from the equilibrium (maximum) value[1] can be, in the first approximation, written in the form of the quadratic function

$$\Delta S = -\frac{1}{2}\sum_{i,k=1}^{n} g_{ik}\alpha_k\alpha_i - \frac{1}{2}\sum_{i,k=1}^{m} h_{ik}\beta_k\beta_i \quad , \tag{3.4}$$

where

$$g_{ik} = \frac{\partial^2 \Delta S}{\partial \alpha_i \partial \alpha_k} \quad , \quad h_{ik} = \frac{\partial^2 \Delta S}{\partial \beta_i \partial \beta_k} \quad .$$

---

[1] ΔS value per unit of volume, i.e., local value of ΔS.

Thermodynamic fluxes are, evidently, determined by the change rate of parameters A
and B, and thermodynamic forces, by deviations of these parameters from their equilibrium values. Let us introduce thermodynamic fluxes in the form

$$J = \frac{d\tilde{\alpha}_i}{dt} \quad , \quad i = 1,2, \ldots, n \quad ,$$

$$I = \frac{d\beta_i}{dt} \quad , \quad i = 1,2, \ldots, m' \quad , \tag{3.5}$$

and thermodynamic forces in the form

$$X_i \equiv \frac{\partial \Delta S}{\partial \alpha_i} = - \sum_{k=1}^{n} g_{ik}\alpha_k \quad , \quad i = 1,2, \ldots, n \quad ,$$

$$Y_i \equiv \frac{\partial \Delta S}{\partial \beta_i} = - \sum_{k=1}^{m} h_{ik}\beta_k \quad , \quad i = 1,2, \ldots, m \quad . \tag{3.6}$$

Fluxes and forces obey phenomenological relations (3.2), phenomenological coefficients satisfy Onsager reciprocal relations

$$L_{ik}^{(\alpha\alpha)} = L_{ki}^{(\alpha\alpha)} \quad (i,k = 1,2, \ldots, n) \quad ,$$

$$L_{ik}^{(\alpha\beta)} = -L_{ki}^{(\beta\alpha)} \quad (i = 1,2, \ldots, n \;\; ; \;\; k = 1,2, \ldots , m) \quad ,$$

$$L_{ik}^{(\beta\beta)} = L_{ki}^{(\beta\beta)} \quad (i,k = 1,2, \ldots, m) \quad . \tag{3.7}$$

In order to prove Onsager's theorem i.e., to deduce (3.7)], it is necessary to
apply the postulate of microscopic reversibility, i.e., of the invariance of the
motion equation of every particle relative to the time sign change. Moreover, it
is necessary to postulate a linear relation between the quenching rate of fluctuation α_i of extensive parameter A_i and the fluctuation value

$$\frac{d\alpha_i}{dt} = M\alpha_i \;\; . \tag{3.8}$$

This means that fluctuations are sufficiently small. Reciprocal relations (3.7)
are thus valid only for the systems comparatively close to equilibrium.

A great part in nonequilibrium thermodynamics is played by the quantity designated as σ and called *entropy production*. This quantity is essentially the rate
of entropy change in the unit volume of the system due to irreversible processes.
It must be emphasized that it is not the total change of system entropy, but only
a fraction of this change that is determined by the irreversible processes going
on within the system. For open systems, the rate of total entropy change, dS/dt, ·
is subdivided into two fractions

$$\frac{dS}{dt} = \frac{d_e S}{dt} + \frac{d_i S}{dt} \quad , \tag{3.9}$$

where $d_e S/dt$ is the entropy flux due to the matter and energy exchange *between the system and the environment*, and $d_i S/dt \equiv P = \int \sigma dV$ is the total entropy production due to the irreversible processes *within* the system.

As it was already mentioned, the entropy production, σ, is the formation of entropy within the system per unit volume and per unit time, i.e., the intensity of the entropy source within the system. If we turn to the local formulae not only for $d_i S/dt$, but for other terms in (3.9) as well, we obtain the following expressions for the second law of thermodynamics in open systems

$$\frac{d\rho s}{dt} = - \text{div } \vec{J}_{s,tot} + \dot{\sigma} \tag{3.10}$$

$$\sigma \geq 0 \quad . \tag{3.11}$$

Here s denotes the entropy per unit mass, ρ density (ρs entropy per unit volume), $\vec{J}_{s,tot}$ total entropy flux per unit area per unit time (3.10) is the equation of entropy balance in an open system with entropy source of σ intensity obeying (3.11). For isolated systems, $\vec{J}_s = 0$, and (3.10,11) are transformed into the conventional formulae of the second law of thermodynamics.

When proving (3.10,11) it was, practically, postulated that the rule $d_i S \geq 0$ is valied for infinitesimal parts of the system. This is more true the closer to equilibrium the system is. The farther from equilibrium the system is, the greater the probability of local entropy fluctuations contradicting (3.11).

From (3.4) we obtain

$$\sigma = \frac{d\Delta S}{dt} = - \sum_{i,k=1}^{n} g_{ik}\alpha_k \frac{d\alpha_i}{dt} - \sum_{i,k=1}^{m} h_{ik}\beta_k \frac{d\beta_i}{dt} \quad . \tag{3.12}$$

Introducing (3.5,6) into (3.9), we find

$$\sigma = \sum_{i=1}^{n} J_i X_i + \sum_{i=1}^{m} I_i Y_i \quad . \tag{3.13}$$

Thus, entropy production is the bilinear function of coupled fluxes and forces. Equation (3.13) allows one to choose the right expressions for irreversible thermodynamic fluxes and forces.

3.3 Stationary States

Let us consider an open system having n independent forces[2], which, according to (3.2), lead to n fluxes

$$J_i = \sum_{k=1}^{n} L_{ik}X_k \quad (i = 1,2, \ldots, n) \quad .$$

The entropy production σ is then

$$\sigma = \sum_{i=1}^{n} J_i X_i \quad . \tag{3.13a}$$

It means that energy $T\sigma$ dissipates in the unit volume within the system per unit time. Introducing (3.2) into (3.13a) and taking (3.11) into account, we obtain

$$\sigma = \sum_{i,k=1}^{n} L_{ik}X_i X_k \geq 0 \quad . \tag{3.14}$$

Let j forces among n be held constant, due to interaction with the environment (for instance, the temperature gradient can be held constant by means of heat supply to the system boundary). What is the state with minimal σ value? In order to find this state, it is necessary to find the extremum of σ with respect to remaining n-j forces that are not held constant

$$\frac{\partial \sigma}{\partial X_i} = 0 \quad (i = j + 1, j + 2, \ldots, n) \quad . \tag{3.15}$$

Taking (3.14) into account, we have

$$\sum_{k=1}^{n} (L_{ik} + L_{ki})X_k = 0 \quad (i = j + 1, j + 2, \ldots, n) \quad . \tag{3.16}$$

According to Onsager reciprocal relations

$$L_{ik} = L_{ki} \tag{3.17}$$

and (3.16) is now

$$2 \sum_{k=1}^{n} L_{ik}X_k = 0 \quad (i = j + 1, j + 2, \ldots, n) \quad . \tag{3.18}$$

The sum here is just J_i [see (3.2)]. Therefore, in the state of minimum entropy production

$$J_i = 0 \quad (i = j + 1, j + 2, \ldots, n) \quad . \tag{3.19}$$

---

[2]For the sake of simplicity, we shall not subdivide these forces into two kinds now, but designate them by one symbol X.

Thus, if a system has n independent forces X_1, X_2, ..., X_n, j of which are held constant (X_1, X_2, ..., X_j = const.), then, in the state with minimum entropy production σ, the fluxes with i = j + 1, ..., n dissappear.

This theorem was proved by Prigogine and named after him. In the state with minimum entropy production, all forces are constant: X_1, X_2, ..., X_j are held constant (but arbitrary), and X_{j+1}, ..., X_n are roots of n-j linear equations (3.18) with n-j unknown quantities. Such a state is called a stationary state of jth order. If j is zero, then in the state with minimum entropy production all fluxes and forces vanish and σ also becomes zero. Thus, the stationary state of the zero order is just the state of thermodynamic equilibrium.

Let us see now, what will be the behavior of the system after the disturbance of the stationary state of jth order with minimum entropy production. We shall now designate the symbols of all forces and fluxes in the stationary state with a right upper zero index. Let m be one of the numbers j + 1, j + 2, ..., n, i.e., the force X_m is not held constant from the outside. Let us add δX_m to this force, keeping the remaining forces unchanged

$$X_m = X_m^0 + \delta X_m \quad (j + 1 \leq m \leq n) \tag{3.20}$$

$$X_j = X_j^0 \quad (j = 1, 2, \ldots, m - 1, m + 1, \ldots, n) \quad . \tag{3.21}$$

Evidently the new value of the flux J_m is

$$J_m = J_m^0 + L_{mm}\delta X_m$$

and since, in accordance with the Prigogine theorem, $J_m^0 = 0$, we have

$$J_m = L_{mm}\delta X_m \quad . \tag{3.22}$$

Now, after the disturbance, σ will be

$$\sigma = \sum_{i=1}^{n} J_i X_i$$

$$= \sum_{i,j=1}^{n} L_{ij} X_i^0 X_j^0 + \sum_{i=1}^{n} (L_{im} + L_{mi}) X_i \delta X_m + L_{mm}(\delta X_m)^2 \quad . \tag{3.23}$$

Here, the first sum is σ^0, and the second is identically zero (put δX_m beyond the sum symbol and use the reciprocal relation; this sum is just $2\delta X_m J_m^0$, but $J_m^0 = 0$ according to the Prigogine theorem). Therefore

$$\sigma = \sigma^0 + L_{mm}(\delta X_m)^2 \quad . \tag{3.24}$$

Since σ^0 is the minimum entropy production, we have

$$L_{mm}(\delta X_m)^2 > 0 \tag{3.25}$$

and [see (3.22)]

$$J_m \delta X_m > 0 \quad . \tag{3.26}$$

The last inequality formulates the Le Chatelier principle for nonequilibrium stationary states. Indeed, the disturbance δX_m results in the flux J_m with the same sign, i.e., decreasing the disturbance. For instance, the increase of grad T will lead to the appearance of a heat flux along the temperature gradient, diminishing this gradient. So, the system, after deviation from the stationary state with minimal entropy production, will spontaneously return to this state.

It is, thus, clear that the existing nonequilibrium thermodynamics is based on linear phenomenological equations and Onsager reciprocal relations. It means that conclusions obtained are only valid for systems close to equilibrium. It is only for such systems that the use of conventional thermodynamic parameters and functions is more or less justified; only for them stable equilibrium states of conventional thermodynamics are replaced by stable stationary states, and, instead of thermodynamic potentials and entropy, approaching constant and extreme values, there appears a new quantity: the entropy production, approaching, in given conditions, a fixed minimum value. In the region of its validity, nonequilibrium thermodynamics is very useful. The ideas and methods of nonequilibrium thermodynamics are especially applicable in the study of various transfer phenomena, including those in biological systems.

All the talk about the specific invalidity of the laws of equilibrium and nonequilibrium thermodynamics for biological systems, which has become quite common in recent years, is due to a misunderstanding. When we are dealing with transitions between states that, within the time intervals of interest to us, can be with good approximation considered as equilibrium states, or when we have practically equilibrated the system parts, living objects obey very well the laws of equilibrium thermodynamics. When the deviation from the equilibrium state is comparatively small (Onsager region), living matter and the processes taking place in living objects satisfy the laws of nonequilibrium thermodynamics. For processes associated with considerable deviations from equilibrium, thermodynamic criteria are meaningless both for the dead and the living nature.

3.4 Dissipative Structures

During the last decade Prigogine and co-workers have been developing the thermodynamics of systems far from equilibrium, where relations (3.2,7) do not hold true. The authors have obtained certain general results and used them to describe certain biochemical processes. In this paragraph a short summary of these results will be given and their significance and prospects discussed. We chiefly refer here to the papers by GLANSDORFF and PRIGOGINE [3.10], and PRIGOGINE [3.11].

There are two types of structures in macroscopic physics: equilibrium and dissipative ones. A crystal can be taken as a representative of equilibrium structures, stable without matter and energy exchange with environment. The ordering of an equilibrium structure depends on the temperature: the lower the temperature, the greater the contribution of intrinsic energy U into the Helmholtz free energy

$$F = U - TS \ . \tag{3.27}$$

Thus, with decreasing temperature, increasingly ordered equilibrium structures corresponding to low entropy values become possible.

The equilibrium state of a system with fixed values of parameters is unique: it does not depend on the system's history. The state of the system is determined by parameters whose differentials are ordinary; there exist potentials, approaching, at equilibrium, extreme values. In other words, the equilibrium state is stable. When deviations are not too great, the system spontaneously returns to the state of equilibrium.

Dissipative structures can exist only due to energy and matter exchange with environment. The dissipation of energy provided from the outside ensures that an ordered structure with an entropy value lower than the equilibrium one is maintained. It was shown in the preceding section that, in the region where linear phenomenological relations between fluxes and forces are valid (i.e., not too far from equilibrium), dissipative structures are just the stationary states of open systems. Instead of the entropy and free energy of equilibrium systems, there appears the entropy production, σ, which has the character of an ordinary differential. Any small deviation from the stationary state due to a fluctuation leads to excess entropy production $\delta\sigma > 0$, and therefore stationary states are stable [Le Chatelier principle; see (3.26)].

If deviations from equilibrium are great and the constraints increase, dissipative structures of a new kind can arise, associated with the initiation of macroscopic motion. The Bénard problem in hydrodynamics can serve as an example. If a liquid layer is heated from below, a temperature gradient and, consequently, a density gradient against gravity will be established. This nonequilibrium stationary state (dissipative structure of the first type) is stable until grad T reaches a certain value after which the system undergoes a sudden transition into the state with the convectional macroscopic motion of a liquid. At grad T exceeding the critical value for the dissipative structure of the first type, the quantity $\delta \int d\sigma$ becomes negative, and a sudden transition takes place as a result of an infinitesimal fluctuation. In the dissipative structures of the second type, part of the system entropy decrease (relative to the equilibrium value) is accounted for by the kinetic energy of the moving liquid

$$\Delta S = -E_{kin}/T \ . \tag{3.28}$$

The appearance of the dissipative structures of the second type is, thus, connected with the instability of the stationary state, i.e., lies beyond the validity boundaries of the Prigogine theorem. The structure arisen may, nevertheless, be in principle characterized by the minimum value of a certain quantity that is an ordinary differential and has the significance of general entropy production. It appears to be true, for instance, in the case of Bênard problem.

The increment of entropy production can be subdivided into two parts

$$d\sigma = d_x\sigma + d_y\sigma \quad , \tag{3.29}$$

where

$$d_x\sigma = \sum_i J_i dX_i \quad , \quad d_y\sigma = \sum_i X_i dJ_i \tag{3.30}$$

see (3.13,13a) . The total entropy production is

$$P = \int \sigma dV \tag{3.31}$$

and, as in (3.29),

$$dP = d_xP + d_yP \quad . \tag{3.32}$$

GLANSDORFF and PRIGOGINE have shown a part of the entropy production to approach the minimum value, even in the absence of a linear relation[3] between the fluxes J_i and forces, i.e.,

$$d_xP = \int dV \sum_i J_i dX_i \leq 0 \quad . \tag{3.33}$$

In the case of macroscopic mechanical processes, i.e., for the dissipative structures of the second type, the quantity

$$d\Phi = \int dV \sum_i J_i' dX_i' \leq 0 \tag{3.34}$$

is introduced. Here the fluxes J_i' and forces X_i' are given with mechanical processes taken into account. In the case of the Bênard problem, $d\Phi$ is an ordinary differential, (3.34) can be integrated, and the system is determined by the extreme value of a certain potential.

However, in the case of the dissipative structures of the second type, it is not always possible to find a parameter with the properties of an ordinary differential. In a chemical reaction system, for example, a dissipative structure of the second type can arise, when, in the macroscopic volume of an open system, concentrations of certain reagents and rates of certain reactions undergo periodic changes. For chemical open systems far from equilibrium it was shown that, in the

---

[3]In the linear region $d_xP = d_yP = dP/2$ and a conventional stationary state with $dP \leq 0$ arises.

general case, there is no quantity of the (3.34) type that would be an ordinary differential. It means, that far from thermodynamic equilibrium, in the general case, there do not exist any quantities of the same kind as entropy production (for the Onsager region of nonequilibrium thermodynamics) or as free energy (for closed systems), having the properties of a potential and determining the state of the system. As a matter of fact, it means that thermodynamic approach to such systems is senseless. Indeed, problems of this kind are being successfully solved with the help of the methods of hydrodynamics or chemical kinetics, and the introduction of rather vague "local potentials" and the attempts to expand the thermodynamic approach beyond the validity region of Onsager relations in applications to concrete systems have not led to nontrivial results.

KEIZER and FOX [3.12] have recently discussed the Glansdorff-Prigogine criterion for the stability of stationary states as applied to open systems of chemical reactions. A detailed analysis, based on the rigorous Liapunov criterion for function stability, has shown that the Glansdorff-Prigogine criterion (i.e., the requirement of the minimum value of the entropy second derivative) loses its sense and usefulness far from equilibrium. The authors have come to the conclusion that this criterion "works" only near the full thermodynamic equilibrium. Analysis of the stability of states far from equilibrium requires not only the thermodynamics, but the kinetic approach as well.

Returning to the dissipative structures of the first and second types, it is appropriate to inquire into their significance in biology. Putting aside the question of the significance of oscillatory chemical processes in biochemical regulation, I cannot refrain from expressing my doubt concerning the prospects of this direction of nonequilibrium thermodynamics (naturally, I speak about biology only). One may, of course, say that the structure of any living system (from cell to multicellular organism) is, after all, maintained by the influx of matter and energy from the outside. In this sense, living systems are dissipative structures. It is difficult, however, to derive anything more from this statement.

The meaningful macroscopic ordering of biological structures does not arise due to the increase of certain parameters of a system above their critical values. These structures are built according to a program like complicated architectural structures, the meaningful information created during many billions of years of chemical and biological evolution being used. Of course, living structures are out of equilibrium at all levels of organization, and during their life this nonequilibrium state is maintained due to the dissipation of energy which enters from the outside. We can say, for instance, that a lathe is maintained in the nonequilibrium state with the help of regular lubrication and replacement of worn-our parts, etc. Biological systems, arising according to a program, are, however, stabilized mainly by kinetic means: transition to the equilibrium state, destruction of the already existing structure requires that a high potential barrier should be overcome.

3.5 Oscillatory Phenomena in Chemistry and Biochemistry

There is an ancient legend. At the city gate the sentries were stopping travellers and asking them: "What have you come here for?". If the traveller answered truthfully, he was stabbed, the liars were hanged. Once a clever man answered: "I have come in order to be hanged". And then an oscillatory process started. If he told the truth he was to be stabbed; but then he would turn out to be a liar and should be hanged; but then he would turn out to have told the truth ... . The sentries were oscillating between two decisions, unable to arrive at any one of them. In this paradox the consequence of a statement contradicts the statement itself.

The origin of the oscillatory chemical reaction is quite similar. In order to observe sustained oscillations of the concentrations of intermediate, resulting from chemical reactions in an open system, it is necessary that at one stage of the process compounds able to inhibit the preceding stages should appear. The result of a process must contradict the process itself. It is necessary to have a negative feedback[4].

The problem of the existence of homogeneous oscillatory chemical systems (where the oscillatory regime is determined by the parameters of chemical reactions only) has a long history. Theoretical predictions of LOTKA [3.13,14,15], VOLTERRA [3.16], and FRANK-KAMENETZKY [3.17,18] for many years were not experimentally verified. The first homogeneous oscillatory chemical reaction appears to have been observed by BELOUSOV [3.19]. However, the first complete and rigorous proof of the existence of a large class of oscillatory homogeneous redox reactions was produced by ZHABOTINSKY and co-workers [3.20-26]. They studied also, for the first time, the space effects in self-maintained chemical oscillatory systems, which can be observed when the diffusion and convectional averaging processes are slower than the chemical ones. They observed mobile and standing concentration waves in a two-dimensional system, which led to the appearance of complicated dynamic or static spatial structures. It is possible that these phenomena will be useful in the investigation of various important biophysical problems (excitation transfer, regulation, etc.).

The intervals of parameter values, that allow self-sustained oscillations to be formed in a system of chemical reactions with negative and positive feedbacks, are so broad, that in the living cell, with its complicated scheme of enzymatic processes self-sustained oscillations just cannot fail to appear. Indeed, many important experimental and theoretical papers have been recently published on the subject (see, e.g., [3.27]). Oscillatory biochemical processes were for the first time observed in [3.35,36]. A systematic experimental study of intracellular os-

---

[4]To be sure, in order to have a steady oscillatory process you need a positive feedback as well.

cillatory biochemical processes was carried out by CHANCE and co-workers [3.37-39]. One cannot doubt that oscillatory processes of a purely chemical origin can occur within the cell. These processes may play an important regulatory role, because they are easily able to tune to the new regime, due to the negligible changes of certain key parameters. It is difficult to say anything more at present. Such a "chemical kinetic" approach is, probably, quite reasonable in studying the complicated schemes of the Jacob and Monod type (see [3.40]), or glycolysis [3.41], etc., if we are not interested in the physical mechanisms of the elementary steps of these schemes.

It is now appropriate to say a few words about the currently rather fashionable monograph of GOODWIN [3.42]. The author has developed the so-called talandic[5] statistical mechanics and thermodynamics based on the idea of chemical oscillators determining the cell behavior. For instance, "talandic temperature" is the measure of the system's deviation from a stationary state due to biochemical oscillations. The lower the talandic temperature and the smaller the amplitude of oscillations, the more "linear" is the system's behavior. In talandic statistical mechanics the part of microscopic variables (the coordinates and moments in conventional statistical mechanics) is played by the concentrations of biochemical components. GOODWIN's theory is ingenious and beautiful. I think, however, that the ideas it is based on are not quite adequate for the object (the cell). I do not think that oscillatory chemical kinetics, in spite of the important part the periodic chemical processes may play in the metabolism and biosynthesis regulation, could be used as the foundation of the general cell theory. It remains obscure if there is anything new except the language, that GOODWIN's approach can bring for the description and understanding of cell properties. It is possible, however, that the skeleton of the theory developed by GOODWIN is something to be kept in storage and will be used in the future after the acquisition of a new physical meaning[6].

3.6 The Problems

It is necessary to use Onsager nonequilibrium thermodynamics in investigations of many important biological processes within the cell or organism. It does not introduce any specific thermodynamic problems associated with particular properties of biological objects. If thermodynamics is spread beyond the validity region of reciprocal relations, it ceases to be thermodynamics (i.e., loses all the advantages of the general thermodynamic approach) and turns into chemical kinetics, hydrody-

---

[5]After the Greek "ταλανδισ".
[6]See also the discussion of Goodwin's theory in [3.2].

namics or some other well-known branch of science. Even if it were possible to develop consistent thermodynamics for regions, where linear phenomenological relations do not hold true (it has not been done so far), it is extremely doubtful, from the author's point of view, that such a theory would be very useful for us to understand the peculiarities of the structure and functioning of biological systems.

4. On the Statistical Physics of Biopolymers

4.1 Where does Mechanics Begin?

It was often emphasized above that biological systems are constructions, i.e.,
structures intended for definite purposes[1]. If, to describe a cell, we should have
to choose between two extreme models — clock mechanism and a homogenous chemical
reaction in the gas phase — the choice would be obvious; a cell is much closer to
a mechanical device than to a pure statistical system. What is the principal difference between statistical and mechanical systems? In order to answer this
question, let us consider a machine for the functioning of which both types of
systems are necessary.

Let us take, for instance, a combustion engine. In order to describe the properties of the fuel mixture and the laws governing the corresponding chemical reactions after their initiation, the statistical approach is necessary. The essential
characteristics of this part of the engine are such macroscopic parameters as
temperature and pressure, the estimation of which requires the averaging throughout the whole phase space, the statistical treatment of an immense number of degrees of freedom of molecules, atoms, ions, and radicals contained in the fuel
mixture, i.e., it requires the statistical approach. The description of the functioning of other engine components: cylinders, pistons, valves, etc., requires an
entirely different, mechanical approach. These engine parts also contain a very
large number of atoms and, generally speaking, are characterized by an immense
number of degrees of freedom; however, for each of these parts there exist only a
few (and may be only one) mechanical degrees of freedom completely determined by
the construction of the system[2].

The cause of the appearance of such specific degrees of freedom lies in the
fact that, for mechanical systems, only a comparatively small number of regions
in the phase space of possible microscopic states are allowed. It must be clearly
understood that this is so not because certain regions of phase space correspond
to excessively high energy. These regions are kinetically unattainable because high

---

[1]About the teleological approach to biology see Sect.2.4.
[2]The rise of mechanical degrees of freedom transforms a construction into a machine.

potential barriers separate them from the regions determined by the pattern of rigid bonds within the system, i.e., by the system construction. For instance, the point in the phase space, corresponding to the localization of piston atoms outside the cylinder, may be energetically nondistinguishable from the point corresponding to the piston's normal position. However, if the engine has been designed correctly, we may disregard the first phase point altogether, because for it to be achieved either the piston or the cylinder must be destroyed and this can happen only after a time interval immeasurably greater than the characteristic time of the functioning of a mechanical system. A construction is characterized by a strictly limited multitude of allowed phase space points, uses but an insignificant fraction of Gibbs canonical ensemble, and, in this sense, is essentially out of equilibrium.

This nonequilibrium state is, however, principally different from the open states, discussed in the preceding chapter, which are called forth by a continuous dissipation of energy coming from the outside. As distinct from such thermodynamic nonequilibrium, the construction nonequilibrium has kinetic nature and is due to the fact that we consider these systems within time intervals much shorter than those required for relaxation to the true statistical equilibrium. Constructions are "frozen" in the nonequilibrium, i.e., ordered, state. In the first chapter we have seen that the ordering of biological construction has meaning. This meaning arises in the course of biological evolution. Here we shall not touch upon the question of the meaning of biological structures, i.e., we shall not consider their functioning. Using most simple examples, we shall try to clarify the consequences of kinetic nonequilibrium in these systems.

A cell is undoubtedly closer to a construction than to a statistical system. It is appropriate to raise the question: at what level of organization does the construction become essential, where does the mechanics begin?

As far as I know, CHURGIN et al. were the first to state that it begins even at the level of individual macromolecules [4.1]. We shall return to their work in Chap.6 when discussing the physical aspects of enzymatic catalysis. We shall see that it is difficult to agree with some of the authors' statements, but the fundamental significance of this work is obvious.

It was LIFSHITZ [4.2,3] who introduced a quantitative description of biopolymer structures as partially nonequilibrium systems (in the kinetic sense stated above). The next section will be completely devoted to his work[3].

---

[3]A rather similar approach to the statistical mechanics of a polymer chain was applied by EDWARDS [4.4], whose works are reviewed in [4.5].

4.2 Statistical Physics of a Linear Homopolymer

Let us begin our discussion with a long flexible polymer chain containing N identical monomer links. N is sufficiently large for the chain to be treated as a statistical system and its possible configurational states as macroscopic phases of this system. Monomers are connected by strong covalent bonds, whose energy E^0 is so great that its breaking frequencies $\omega \sim \exp(-E^0/kT)$ satisfy the inequality

$$\omega\tau \ll 1 \tag{4.1}$$

where τ denotes the characteristic time of polymer chain configurational relaxation. Inequality (4.1) means that the primary structure of the polymer chain is kinetically stabilized, and continuously acting correlative relations exist between the position of neighboring links. In fact, (4.1) implies that during the time intervals ($\sim\tau$), which are important here, the system is out of equilibrium (kinetically) and has a linear memory.

Taking into consideration the linear memory leads to nontrivial consequences even for a free polymer chain (without secondary side bonds). It is well known (see [4.6]) that such a chain, due to bending fluctuations at high N values, folds into a coil with radius $R_N \approx a\sqrt{N}$ where a is, by its order of magnitude, equal to the interlink distance. Due to the linear memory, partition function Z_{free} is expressed in the form of a product of correlation functions $g_j = g(\underline{x}_j, \underline{x}_{j+1})$ between neighboring links

$$Z_{free}(\underline{x}_1, \ldots, \underline{x}_N) = \prod_j g_j \quad . \tag{4.2}$$

LIFSHITZ defined the correlation function through the vector difference

$$g_j = g(\underline{x}_{j+1} - \underline{x}_j) = g(\underline{y}_j) , \quad \int g \, d^3\underline{y} = 1 \tag{4.3}$$

where $\underline{x}_j (j = 1, 2, \ldots, N)$ are the generalized link coordinates, and introduced the link density at point $\underline{x}$

$$n(\underline{x}) = \sum_{j=1}^{N} \delta(\underline{x} - \underline{x}_j) \quad . \tag{4.4}$$

He then estimated the density correlation—a quantity connecting the density change at one point within the coil with the density change at another point. The correlation radius seems to be of the same order of magnitude as that of coil dimensions. It means that coil density cannot be regarded as a thermodynamic parameter of the system: it is impossible to obtain the constant average density value by averaging the density throughout the coil. Density fluctuations at different points within the coil are interdependent, and the averaging must be carried out through time but not through space. Due to the density correlation (i.e., due to the linear memory that introduce limitations to the allowed regions of phase space) the coil must undergo macroscopic pulsation with characteristic time

$$\tau \sim N^2 \qquad\qquad (4.5)$$

and with pulsation amplitude of the same order of magnitude as the density itself.

This result is, essentially, connected with the free flexible chain approximation and with all types of interactions being neglected, except the principal ones which call forth the linear memory. In actual fact, volume interactions between distant (along the main chain) monomer links, brought together due to chain folding, cannot be neglected. These side bonds determine, to a considerable degree, the resulting configuration of a macromolecule. The interaction of the polymer chain with the environment (e.g., with the solvent) can also be quite important. The presence of the hydrophobic polymer chain residues on the coil surface results in the disruption of hydrogen-bonded water structure, i.e., in the so-called hydrophobic interaction, pressing the coil together. Therefore, as a second approximation, LIFSHITZ considered a coil, formed by a homopolymer flexible chain, in an external pressing potential field $u(\underline{x})$. This stage of his procedure is necessary for the subsequent treatment of volume interactions.

Every link has now an energy determined by the space position of the link. Let us designate the energy (in T units) of the link having the coordinate $\underline{x}$, by

$$\varphi(\vec{x}) = u(\underline{x})/T \quad . \qquad\qquad (4.6)$$

The energy of the whole system in a fixed configuration Γ (i.e., with fixed coordinates $\vec{x}_j$ of all links[4]) is now

$$E = \sum_{j=1}^{N} \varphi(\underline{x}_j) = \int n(\underline{x})\varphi(\underline{x})d^3\underline{x} \quad . \qquad\qquad (4.7)$$

Free energy $F(x)$(in T units) of the system in the field, from the free coil level, may be written in the form

$$F_1/T \equiv F \equiv F(\varphi) - F(0) = -\ln Z_N \quad , \qquad\qquad (4.8)$$

where F_1 is the free energy in conventional units. In the partition function Z_N, the external field, as well as linear memory, are accounted for

$$Z_N = \int \exp\left\{-\sum \varphi(\underline{x}_j)\right\} \prod_j g(\underline{y}_j)d\Gamma$$
$$d\Gamma = \prod_j d^3\underline{x}_j = \prod_j d^3\underline{y}_j \quad . \qquad\qquad (4.9)$$

We can now write obvious expressions for the equilibrium link density $\bar{n}(\underline{x}_j)$, energy E (in T units) and entropy S of the system

---

[4]For the sake of simplification, we assume that the state of a link is fully determined by its coordinate, and the influence of possible orientations is negligible.

$$\bar{n}(\underline{x}_j) = \frac{\delta F}{\delta \varphi} \tag{4.10}$$

$$E = \frac{E_1}{T} \equiv E\{\varphi\} - E\{0\} = \int \varphi(\underline{x})\bar{n}(\underline{x})d^3\underline{x} = \int \varphi(\bar{x}) \frac{\delta F}{\delta \varphi} d^3\underline{x} \tag{4.11}$$

$$S = S\{\varphi\} - S\{0\} = E - F = \int \varphi(\underline{x})\bar{n}(\underline{x})d^3\underline{x} - F \quad , \tag{4.12}$$

where E_1 is the energy in conventional units. If N is high enough, the estimation of Z_N is, in the end, reduced to the solution of the equation[5]

$$\int g(\underline{x} - \underline{x}')\Psi(\underline{x}')d^3\underline{x}' = \exp(\varphi - \lambda)\Psi(\underline{x}) \tag{4.13}$$

where Ψ is a function determined by means of equality

$$\Psi^2(\underline{x}) = \frac{n(\underline{x})}{N} \exp[-\varphi(\underline{x})] \equiv \rho(\underline{x})\exp[-\varphi(\underline{x})] \quad , \tag{4.14}$$

$\rho(\underline{x})$ is the link density per one link at $\underline{x}$, λ is the energy parameter of the system related to its free energy

$$Z_N \approx \exp(-\lambda N) \quad , \tag{4.15}$$

$$F = N\lambda = N\lambda\{\varphi\} \quad . \tag{4.16}$$

If function $\varphi(\underline{x})$ is sufficiently smooth, i.e., varies but slowly at distances comparable with the interlink spacing, and correlation function $g(\underline{x})$ may be considered spherically symmetric, (4.13) then turns into

$$a^2\Delta\Psi + [1 - \exp(\varphi - \lambda)]\Psi = 0 \quad . \tag{4.17}$$

The solutions of this equation, i.e., the "fundamental functions" and the allowed "proper values" λ, depend on $\varphi(\underline{x})$. If the field vanishes at infinity, and in the coil region $\varphi(\underline{x})$ represents a potential well deep enough to hold the coil in the form of a dense globule, there appear discrete proper values λ. With a given potential field this takes place at temperatures lower than a certain critical temperature θ_c. At higher temperatures, there is no free energy quantization and the spectrum of λ is continuous. The boundary of the continuous spectrum corresponds to $\lambda_c = 0$. If the well is broad enough, the ground level lies near the well bottom, i.e.,

$$|\varphi(\underline{x}) - \lambda| \ll 1 \quad . \tag{4.18}$$

In this case, instead of (4.17), we have

$$a^2\Delta\Psi + (\lambda - \varphi)\Psi = 0 \tag{4.19}$$

---

[5]For detailed treatment see [4.2].

formally identical to Schrödinger equation.

This equation can be solved for various concrete cases. Thus, for a deep spherically symmetric well $\varphi(\underline{r}) = \varphi_0 + dr^2/2$ and the ground level corresponds to

$$\lambda = \varphi_0 + \frac{3}{2} \alpha^{3/2} \quad . \tag{4.20}$$

At $T > \theta_c$ there are no discrete levels, and in the course of temperature decreases there appears the first discrete level, $\lambda_c = \lambda\{\varphi_c\} = \lambda\{U(\underline{x})/\theta_c\}$. For T close to θ_e we have

$$\lambda\left\{\frac{U}{T}\right\} - \lambda\left\{\frac{U}{\theta_c}\right\} \sim (T - \theta_c)^2 \quad . \tag{4.21}$$

At θ_c a phase transition of the second kind is, thus, realized

$$\frac{\frac{F_T}{T}}{N} = \lambda_T = \begin{cases} A(T - \theta_c)^2 , & T < \theta_c \\ 0 , & T > \theta_c \end{cases} \tag{4.22}$$

(here F_T coincides with F_1).

Analysis of the correlations of density fluctuations shows that, at $T < \theta_c$, if the well is deep enough to hold the coil, density $\rho(\underline{x}) \sim \psi^2 \exp(\varphi)$ becomes a thermodynamic quantity and the process of its space averaging becomes meaningful again.

In an actual macromolecule, in addition to the interaction between neighboring links, that is accounted for by g_j correlation, there exists a supplementary volume interaction U between the links localized far from each other along the chain but brought together due to chain folding. The system energy is now the sum of pairwise interaction energies u_{ik} between the converging links

$$U(\underline{x}_1, \ldots, \underline{x}_N) = \frac{1}{2} \sum u_{ik} ; \quad u_{ik} = U(\underline{x}_i - \underline{x}_k) \quad . \tag{4.23}$$

In the general case, when there is volume interaction U, and external potential field φ as well, equations analogous to (4.13,17,19) can be derived, but, instead of $\varphi(\underline{x})$, one must introduce the density dependent self-consistent field $\Phi(n,\underline{x})$

$$\Phi(n,\underline{x}) = \varphi(\underline{x}) + \mu_0(n) - \ln n \tag{4.24}$$

where $\mu_0(n)$ is the chemical potential of the link in the system having the local density n.

Now, about the application of the term "chemical potential" in this case [4.7]. All the pecularities of polymer chain statistics are determined by the fact that, due to linear memory, monomers prove to be correlated at the distances of r_{cor} ($r_{cor} \gg a$). For this reason, there is no local density distribution within the globule. However, in a first approximation (relative to parameter a/r_{cor}), volume interaction can be accounted for by means of effective chemical potential $\mu^*(n,T) = \mu(n,T) - T \ln n$ where $\mu(n,T)$ is the chemical potential of the same group of disrupted links with the same volume interaction, density n, and temperature T. Thus, chemical potential is introduced in the same way as for a condensed phase without linear memory, because from a system with linear memory a volume is separated where local distances between the interacting links are much lower than the distances between the same links along the chain (the model of "interacting beads on a flexible fiber").

The influence of the solvent upon the density distribution can be accounted for by the effective chemical potential, $\mu^*(n)$, dependence on the solvent state. For instance, $\mu^*(n)$ depends on the pressure P within the solvent. This dependence is mainly due to the fact that, in the course of a dense globule formation, the solvent volume changes by the quantity ΔV (the solvent is forced out by polymer links, because solvent molecules are not able to penetrate into the dense globule). Therefore, during the globule formation, the chain performs work $P\Delta V$. In this case, the polymer chain state is determined not by the minimum of Helmholtz free energy, but by the minimum of the function analogous to Gibbs free energy (to the thermodynamic potential): $G(n) = F(n) + P\Delta V$. In expressions for the effective chemical potential and for the effective self-consistent field (4.24) now appear additional terms, determined by the solvent state (pressure, pH, etc.).

If the external field is absent, and the first term on the right side of (4.24) vanishes, the coil - globule transition may, nevertheless, take place due to the volume attraction forces. To prevent unlimited compression, it is necessary to take into account the forces of repulsion acting at atomic distances. In this case, the globule forms a kind of two-phase system with a dense nucleus, which plays the part of the condensed phase, and a "gaseous" periphery, where the field of volume interaction is too weak to lead to globule formation, and the chain is held by the main bonds only.

In the following discussion we shall again designate the energy characteristics of the system conventionally, i.e., we shall use F/T instead of F, and $\mu_0(n,T)/T$ instead of μ_0.

Let us write out the state equation for a two-phase spherically symmetrical globule for $\varphi(\underline{x}) = 0$

$$\int g(\underline{x} - \underline{x}')\Psi(x')d^3x' = \Psi(\underline{x})\exp[\varphi(n) - \lambda] \quad \text{(a)}$$

$$\Psi^2 = n \exp[\lambda - \Phi(n)] \quad \text{(b)}$$

$$\Phi(n) = \mu_0(n,T)/T - \ln n ; \quad n = n(\underline{r}) \quad \text{(c)}$$

$$n(\infty) = \Psi(\infty) = 0 \quad \text{(d)}$$

$$(4.25)$$

and the normalization condition that fixes the value of λ in the case of globule formation

$$4\pi \int n(r)r^2dr = N \quad . \quad (4.26)$$

40

If density n is low (the "gaseous" phase), the chemical potential, according to conventional thermodynamics, depends on n

$$\mu_0/T = \ln n + \alpha(T)n \qquad (4.27)$$

function $\alpha(T)$ (the second virial coefficient) being positive at high and negative at low temperatures. For a condensed phase (large n values), Φ has a minimum Φ_{min} at a certain n_m. Thus, the self-consistent field $\Phi(n)$, caused by volume interaction, depends on the density in the following way:

$$\Phi(n) = \begin{cases} \alpha(T)n \quad, & \alpha(T)n \ll 1 \quad, \\ \Phi_{min} + \gamma(n - n_m)^2 \quad, & |n - n_m| \ll n_m \quad. \end{cases} \qquad (4.28)$$

Figure (4.1) illustrates the typical shape of $\Phi(n)$ functions.

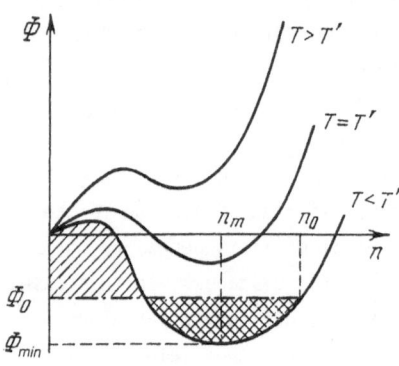

Fig.4.1. Typical shape of $\Phi(n)$ functions (according to [4.2]). T' is the temperature at which the energies of dense globule and of free coil coincide. The horizontal dot-and-dash line for the $\Phi(n)$ function at T < T' must be drawn in such a way that the shaded areas are equal

For large globules, the solution of (4.25) means that practically the whole mass of the globule is localized within the dense nucleus of an almost constant density n_0 ($\frac{4}{3}\pi R^3 n_0 \approx N$). The transition layer and the "gaseous" periphery are rather thin ($\delta R \ll R$). By analogy with (4.17) for each phase across both sides of the thin transition layer one can rewrite (4.25) in the form

$$a^2 \frac{d^2\psi}{dx^2} = \psi(\exp[\Phi(n) - \lambda] - 1) \quad, \quad \psi^2 = n \exp[\lambda - \Phi(n)] \qquad (4.29)$$

where x = r-R, n is now a function of x and the boundary conditions for condensed (I) and "gaseous" (II) phases are

Phase I $n(-\infty) = n_0$

$\qquad\qquad \psi^2(-\infty) = \psi_0^2 = n_0 \exp(\lambda - \Phi_0) \quad,$

Phase II $n(\infty) = \psi^2(\infty) = 0 \quad;$ $\qquad\qquad\qquad\qquad\qquad\qquad$ (4.30)

$n(-\infty)$ designates the link density within the globule[6] far from its boundaries, i.e., at $r \ll R$.

From (4.29,30) we at once obtain

$$\lambda = \Phi(n_0) = \Phi_0 \quad , \quad \psi_0^2 = n_0 \quad . \tag{4.31}$$

Far from the phase boundary (4.25) are the exponential functions

$$\left.\begin{array}{l} n(x) - n_0 \sim \exp(\nu_1 x) \quad (x \to -\infty) \\[2mm] n(x) \sim \exp(\nu_2 x) \quad\quad (x \to \infty) \end{array}\right\} \tag{4.32}$$

where ν_1 and ν_2 are certain functions of n_0.

LIFSHITZ obtained n_0 matching the solutions for ψ_1 and ψ_2 of the exact equation (4.25), but not of its differential form (4.29), which is valid far from the phase boundary. At each point thermodynamic inequality [see (4.25c)]

$$\frac{1}{T} \frac{\partial \mu_0}{\partial n} = \frac{\partial \Phi}{\partial n} + \frac{1}{n} > 0 \tag{4.33}$$

must hold true, and a dense globule must be thermodynamically preferable to a loose coil, i.e.,

$$\int_0^{n_0} \Phi(n)dn < 0 \quad . \tag{4.34}$$

The first integral of (4.29), [taking into account (4.30,31)], is

$$a^2 \left(\frac{d\psi}{dx}\right)^2 = \begin{cases} \int_0^n (\Phi - \Phi_0)dn - g(n) & \text{(Phase I)} \quad , \\[4mm] \int_{n_0}^n (\Phi - \Phi_0)dn - g(n) & \text{(Phase II)} \quad , \end{cases} \tag{4.35}$$

where $g(n) = n[\exp(\Phi - \Phi_0) - 1 - (\Phi - \Phi_0)]$. The difference between the right parts of (4.35) is $\int_0^{n_0}(\Phi - \Phi_0)dn$. Therefore, if we want the solution of (4.29) to be continuous, the equality

$$\int_0^{n_0} (\Phi - \Phi_0)dn = \int_0^{n_0} \Phi(n)dn - n_0\Phi(n_0) = 0 \tag{4.36}$$

must be satisfied.

Hence [see (4.34)],

$$\Phi_0 \equiv \Phi(n_0) < 0 \quad . \tag{4.37}$$

---

[6] Of course, $x = r - R$ cannot be less than $-R$.

Equation (4.36) fixes n_0. There exists a certain temperature T' below which (4.34, 36,37) are satisfied. This temperature is determined by equations

$$\Phi(n_0,T') = 0 \quad , \quad \int_0^{n_0} \Phi(n,T')dn = 0 \quad . \tag{4.38}$$

Moreover, (4.33) must hold true in the whole interval of density values from 0 to n_0, i.e., temperature T must be higher than the critical temperature ($T > T_c$). Here, the word "critical" has the conventional sense: it means the temperature boundary of continuous transition from "gas" to "liquid". Figure 4.1 makes clear how one can find n_0 with the help of $\Phi(n)$ using (4.36).

So, for a sufficiently long homopolymer with volume interaction, the following situations can be realized:

1) $T' > T_c$.

1a) $T > T'$. State equation (4.25) does not have any solutions for a two-phase system.

A globule does not arise. The system behaves as a free chain without volume interaction.

1b) $T_c < T \leq T'$. A globule appears at $T = T'$. Additional free energy arises, equal at $T < T'$ to $F/TV = n_0\Phi(n_0)$. (4.36) is satisfied. The transition at $T = T'$ is a phase transition of the first kind.

1c) $T \leq T_c$. At $T = T_c$ a density jump (phase transition of the second kind) occurs. The condition (4.36) in the nucleus center is, nevertheless, fulfilled.

2) $T' < T_c$.

2a) $T > T'$. As for (1a).

2b) $T \leq T'$. At $T = T'$, due to phase transition of the first kind, there appears a globule with a density jump between the dense nucleus and the "gaseous" periphery.

Hitherto we were dealing mainly with the first paper by LIFSHITZ [4.2]. Recently [4.3,7] LIFSHITZ and GROSBERG have obtained new essential results. A possibility of globule formation without a "gaseous" periphery has been considered, the cases of large and small globules analyzed, and the influence of the solvent taken into account. In the remaining part of this section the most important of these results will be described.

With temperature decrease, link density in the "gaseous" periphery drops as in the case of vapor density above the liquid surface. In the case of a polymer chain, however, the main bonds do not allow the density to become too low. Therefore, at a sufficiently low temperature, even a very long chain forms a globule without the "gaseous" periphery.

Thus, three different phase states of a polymer chain can exist: a coil, a globule with a "gaseous" periphery, a globule without a periphery. These three states are schematically shown in Fig.4.2.

A detailed analysis of the thermodynamics of these states and of transitions between them was performed using the already mentioned model of "interacting beads on a flexible fiber". The cases of extremely large (the volume of the globule's dense nucleus $V_0 = (4/3)\pi R^3$ is considerably greater than the volume of correlation: $V_0 \gg r_{cor}^3$), and extremely small ($V_0 \ll r_{cor}^3$) globules are discussed. The main results can be seen in the phase diagram (Fig.4.3).

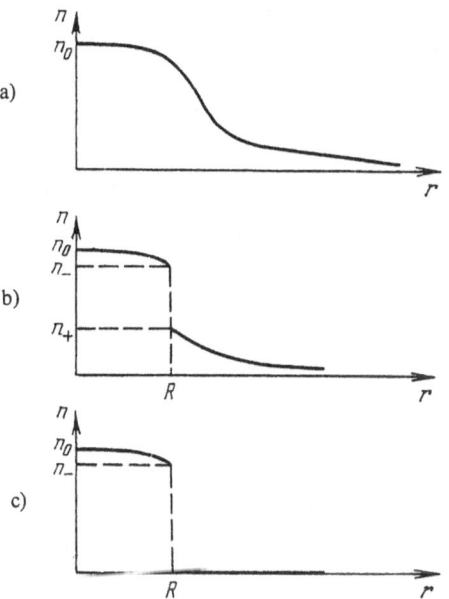

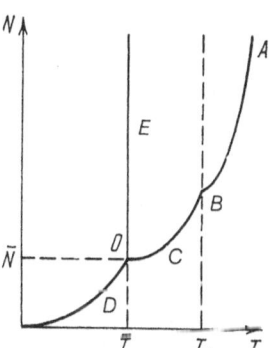

Fig.4.2a-c. Radial density distribution; (a) in a globule with "gaseous" periphery at temperatures higher than the critical temperature of a system with broken links ($T > T_C$); (b) in a globule with "gaseous" periphery at $T < T_C$; (c) in a globule without "gaseous" periphery. n_- and n_+ are the densities inside and outside the dense globule boundary, respectively

Fig.4.3. Phase diagram for the "interacting beads on a flexible fiber" model at P = const. EOD lines separate the region of the globule without a shell, ABCOE, the globule with gas-like shell; ABCOD, the coil

For a large globule (N is sufficiently high), the transformation of a coil into a globule with a gas-like shell is realized as a phase transition of the first kind (line AB). In the neighborhood of this phase transition, where $0 < \tau = T' - T \ll T'$ (T' is the transition temperature for given N), free energy is a linear function of

τ. If $T' > T_c$, the intersecting T_c value (dashed line in phase diagram 4.3) is accompanied by a phase transition of the second kind, due to which density distribution on the boundary between the globule and the shell becomes discontinuous (compare with Fig.4.2b).

For a not sufficiently large globule, $T_c > T'$, and only phase transformations of the first kind: coil → globule with a gas-like shell can be realized (line c in phase diagram).

If the system of main interlinking bonds does not allow the requirements of the local chemical potential continuity on the nucleus boundary to be fulfilled, a globule without a shell appears. In the case of an extremely short chain (but long enough to be regarded as a statistical system!), the globule is always formed without a shell as a result of phase transition of the first kind (line D in phase diagram). For both large and small globules with gas-like shells, in the course of temperature decrease, phase transition of the first kind is realized (line E in phase diagram) as a result of which the gas-like shell "hides" within the globule. The triple point $O(\bar{N}, \bar{T})$ is accounted for by the fact that at this point free energies of both kind of globules (those with and without a shell) vanish. It can be shown that $\bar{T} < T_c$.

In [4.7] the external pressure dependence of a phase diagram is described. At low enough temperatures the formation of a globule without a gas-like shell can always be observed.

The obvious development of the model considered above would be by taking into account not only the coordinates of monomer links, but their orientations as well. Important progress would be the treatment of heteropolymers possessing not only the linear position memory but the information memory as well. A link is characterized now not only by a coordinate or orientation but by the type of monomer also. However, these first steps towards the formulation of the statistical theory of biopolymers, taking into account their most important characteristic — the existence of linear memory, have already brought about very interesting and important results that will surely remain in the more complete theory. New results can be found in [4.9-11].

4.3 On the Statistical Nonequilibrium of Biopolymer Structures at Different Levels of Organization

In the preceding section, while discussing the properties of a unichain homopolymer, we assumed that kinetic nonequilibrium is restricted to the localization of each monomer link at its site along the chain. In the case of a heteropolymer, it should be said that memory concerns only the primary structure. The formation of a coil, or a dense globule with fixed discrete free energy values, or of a two-phase structure with a condensed nucleus and a gas-like pulsating shell, leads, according to the ideas formulated in the preceding section, to the equilibrium state of the system (taking into account, of course, kinetic limitations due to the system of strong bonds along the chain). Thus, in this approximation, the structures of secondary and tertiary biopolymers are at equilibrium and are fully determined (at T < T') by the given nonequilibrium (kinetically) primary structure.

It appears, however, that it is not always only the primary structure that is out of equilibrium. The actual protein molecules with a large enough number of aminoacid residues, capable of realizing an immense quantity of different pairwise interactions, possess, probably, a multitude of near-lying free energy levels, corresponding to various ways of secondary bond formation and separated from one another by high potential barriers. In this case, the coincidence of the unique meaningful (i.e., possessing specific ensymatic, immunological, etc., activity) structure with that corresponding to the absolute free energy minimum for a given aminoacid sequence is highly improbable. Apparently, the native and absolutely equilibrium structures cease to coincide starting from a certain chain length.

This does not mean that the secondary and tertiary structures of biopolymers within the cell are not determined by the genetically fixed primary sequence. If kinetic mechanisms of biosynthesis are fixed, the primary structure completely determines the secondary and tertiary structures, which, however, may be out of equilibrium in the above meaning of these words. Indeed, let us postulate the now accepted scheme of protein biosynthesis in ribosomes, where aminoacid residues brought by the transport RNA are one by one forming peptide bonds with each other according to the instruction written in the messenger RNA, and the ready polypeptide chain is gradually expelled from the ribosome into the cytoplasm. When the growing polypeptide chain, fixed at one side in the ribosome, is long enough, there arises a situation when the transformation of the free coil into a dense globule must take place. One cannot be sure, however, that the lowest of the discrete free energy levels, corresponding to a free chain of this length and composition, will be achieved. It is possible that, at first, a thermodynamically less favorable but kinetically more easily accessible conformation may be formed which would then slowly relax to an equilibrium one if the chain growth terminates and a ready protein molecule leaves the ribosome. If, however, the chain growth proceeds, the packing of the new chain regions will be determined by the already existing "nucleus". After some time, in the chain that has grown longer, the conformation of the initially formed dense globule (its secondary and tertiary structure) may be brought well out of equilibrium. The lowest free energy level of the grown polypeptide chain would correspond to a quite different conformation of the initially synthesized chain regions where the dense globule had been previously built. The transition to a new conformation may, however, be kinetically impossible. Such a transition would require either the breaking of the main chain or the destruction of the whole globule through a gradual disruption (in a strict order) of a multitude of secondary bonds.

Thus, for sufficiently large biopolymer molecules, the "rigid" memory may include not only primary but also secondary and tertiary structures. However, with given mechanisms and time intervals of biosynthesis, the kinetically out of equilibrium states of secondary and tertiary structures are not inconsistent with the fact that they are determined by the primary sequence of monomers, i.e., are genetically programmed.

I, therefore, think that the problem of the degree of equilibrium in secondary and tertiary protein structures (with a fixed primary sequence), is not very important. In actual fact, this problem may be altogether meaningless. The functioning of the most important biopolymers is associated with constantly occurring microchemical changes: addition and dissociation of low-molecular compounds, ionization of acid and alkaline groups, changes in the number of adsorbed ions, etc. Any of these local changes result in the change of the effective field Φ, i.e., may lead to the transformation of an equilibrium conformation into a nonequilibrium one. It is not really important whether the conformation in question is in absolute equilibrium or at some local equilibrium, i.e., has the lowest free energy level with respect to the few neighboring conformations allowed. We shall see in the following chapters that the most important problem is the question of times and mechanisms of system relaxation to a new equilibrium (or quasiequilibrium) state after a rapid change of the effective field Φ.

A discussion of related topics may be found in [4.3,8].

4.4 On Certain Properties of Biopolymer Structures that Can Be Understood in Terms of Their Statistical-Physical Description

We can acknowledge now that the most important characteristic of biopolymers is the existence of a rigid memory, i.e., a kinetic nonequilibrium at the level of the primary structure, and, for sufficiently large molecules, at the level of the secondary and tertiary structures. According to Lifshitz theory, rigid memory and volume interaction leads to the appearance of discrete free energy levels in the dense globule.

In the case of heteropolymers, every such level must, apparently, correspond to a specific conformation, i.e., to the specific way the secondary bonds (hydrogen bonds, disulfide bridges ion pairs) are formed between aminoacid residues remote from each other along the main chain. This specific way determines the secondary and tertiary structures of the macromolecule[7]. (Footnote see p.47)

The existence of these conformational levels separated by energy and entropy barriers from other conformational states, their genetic condionality, and the constancy of kinetic mechanisms of biosynthesis result in a remarkable reproducibility of biopolymers. Indeed, molecular distribution functions of identical (i.e., corresponding to one gene) proteins with respect to various properties are, for instance, extremely sharp, much sharper than the corresponding distribution functions of any synthetic polymer. This reproducibility ensures the functional specificity of biopolymers (enzymatic, immunological, etc.) so necessary for living systems.

At the same time, rather weak local (e.g., the formation of an additional bond with a small molecule or ion) or general (e.g., a small temperature change) perturbations, not strong enough to transform a molecule into a new conformational state (i.e., to change the pattern of secondary bonds which determines the volume interaction), may lead to comparatively great changes in the geometry configuration of a molecule. In this respect, the properties of large molecules, possessing secondary and tertiary structures, are quite different from those of small molecules. For biopolymers, the free energy minimum at the same conformation may be realized at somewhat different positions and orientations of monomers, not connected by secondary bonds with each other. This is favored by the possibility of rotation about the single bonds of the main chain, by not too strict rules concerning the linearity of hydrogen bonds, etc.

The structure of a biopolymer molecule resembles a complicated wire construction, the shape of which may change due to heating or weak local constraints. This specific property of large chain molecules with a secondary and tertiary structure must be taken into account when their chemical interactions with each other and with small molecules are analyzed.

Sufficiently strong perturbations can lead to true conformational transitions with the old bonds being broken and new secondary bonds formed. If discrete free energy levels lie close enough to each other and are separated by comparatively low energy or entropy[8] barriers, one may expect the rise of conformational oscillations of the relaxation origin. These transitions and oscillations may, in some cases, have local character and not involve the whole molecule.

When analyzing all the phenomena associated with configurational and conformational changes of biopolymers, one must bear in mind that these structures, as it was shown above, are, to a considerable extent, mechanical systems with rigid

---

[7]The word *conformation* does not have a definite meaning in the current literature. In this book we use the term "conformation" only to designate the state of a macromolecule with a fixed pattern of secondary bonds. By conformational changes and transitions we shall understand those connected with the changes in this pattern. The geometrical shape of a molecule that determines the distances between the monomers not connected by secondary bonds and can change within the limits of one conformation, will be called a *configuration*, and its changes *configurational* changes.

[8]A low entropy barrier means that for a transition between two states to take place it is necessary to change, one after another, a comparatively small number of weak bonds, or that this transition may be realized in many ways.

memory at various levels of organization, with many regions of phase space unattainable without the destruction of the system. Therefore, one can expect few specific degrees of freedom to exist, and the motion of different parts of the system during conformational and configurational changes must have mechanical character. One can also expect rather large characteristic times of relaxation transitions to the new state of partial equilibrium after fast perturbations.

All the properties of biopolymer structures mentioned above must greatly influence the functioning of these structures.

5. Conformational and Configurational Changes of Biopolymers

5.1 Introductory Remarks

At the end of the preceding chapter we saw what could be the consequences of the fact that most important biopolymers constitute kinetically nonequilibrium systems with memory at different levels of organization and have specific degrees of freedom, i.e., are machines. In order to describe the true motion of such systems in space and time, a considerably smaller number of degrees of freedom must be taken into account than is in principle necessary for their complete description. There are two reasons for such "mechanical behavior" of biopolymer macromolecules: their large size and the possibility of a great number of secondary bonds to be formed, strong enough to fix different conformations (discrete free energy levels, according to [5.1]), and weak enough to allow configurational and conformational transitions under the influence of various perturbations.

It is just the ability to undergo conformational transitions, the ability of motion with the participation of specific mechanical degrees of freedom that is the most important property of biopolymers, to a considerable degree determining their functioning. Many papers (mostly experimental) concerning these transitions have been published.

However, when analyzing their experimental data, the authors, as a rule, do not take into account the above-mentioned specific properties of biologically important macromolecular structures. Therefore, in this chapter, experimental data concerning conformational and configurational biopolymer changes will not only be described, but analyzed all over again (except for the data directly concerning the enzymatic catalysis, which will be discussed in Chap.6).

The greater part of the published papers is devoted to studying the so-called denaturation of biopolymers. This term does not have a definite meaning, and is used to designate various changes of macromolecular properties under the influence of quite different agents. In the course of analyzing the thermodynamic and kinetic aspects of various denaturation processes there arises the general problem of the applicability of ideas and equations, used in the physical chemistry of small molecules, to the description of processes in which biopolymer molecules take part. It is especially the question of the applicability of mass action law, Van't Hoff and

Arrhenius equations and the activated state theory. This question is very important
for many problems which will be discussed in this volume, and, therefore, it will
be subjected to a detailed analysis in this chapter, though the principal conclusions drawn from this analysis (concerning the kinetics and mechanisms of enzyme
reactions and mechanisms of energy transformation within the cell) will be utilized
only in the following chapters.

Considerable attention will be given to thermodynamic and kinetic aspects of
protein conformational and configurational changes observed as a result of local
chemical perturbations (complex formations with small molecules and ions, acquisition or loss of an electron, etc.).

5.2 Biopolymer Denaturation

There exists an immense multitude of original articles, reviews, and special chapters
in textbooks, concerning protein denaturation, and it is impossible even to enumer-
ate them. It may be even unnecessary, because, with the progress of physical methods
of investigation, it becomes more and more evident that there does not exist one definite denaturation process, the properties and mechanism of which are to be established. Considerable macromolecular conformational changes occurring within narrow
intervals of external parameters (temperature, pH, hydrostatic pressure, etc.) are,
probably, only one common characteristic of all the effects accompanying the process
called denaturation. In this sense, we may speak of the cooperativity of denaturation
changes which sometimes resemble phase transitions. Probably, these changes are
often phase transitions (globule - coil) that were considered in Chap.4.

The melting or helix-coil transition in DNA is a denaturation process that has
been thoroughly studied both experimentally and theoretically. The most importance
works are listed in [5.2-8]. Special attention is to be given to the works of
LAZURKIN and co-workers, who, in my opinion, have made the most important contribution to the creation of a consistent concept and to the clarification of the detailed mechanism of double-helical DNA fiber melting. In the case of protein denaturation, there are no and, probably, cannot be any works of this type. The detailed mechanism of melting, of conformational globule-coil transition, is determined by the protein macromolecular specific conformation. In this sense, there must
be as many mechanisms and theories as there are different proteins.

A more or less general approach will be possible only after statistical physics
has been developed for systems with memory at different levels of organization. The
beginning was made by Lifshitz in his works discussed in Chap.4.

For many years a rather primitive approach was applied to the thermodynamic description of denaturation processes. A denaturation was regarded as a conventional
monomolecular chemical reaction, in which compound A (native form of biopolymer) is

transformed into compound B (denatured form). It is generally assumed that this transformation can be carried out as thermodynamically reversible, and that at fixed values of external parameters (temperature, pH, concentration of the agent causing denaturation, etc.) there exists a true equilibrium relationship between A and B concentrations. This has been regarded as a sufficient reason to use the mass action law and Van't Hoff equation for reaction isotherm. Kinetic results are treated by means of the Arrhenius equation and concepts of the theory of absolute rates, using conventional methods for the estimation of activation energy and entropy values. It is only quite recently, due to the development of the high-precision microcalorimetry technique, that direct measurements of the heat effects of denaturation processes (as well as of complex formation between small molecules and proteins, and of association of macromolecules) began to be carried out. The comparison of equilibrium and calorimetric results compels us to doubt the validity of the former. Moreover, many results of equilibrium measurements arouse by themselves a sceptical attitude. Indeed, heat effects and reaction entropy values (ΔH and ΔS), determined according to Van't Hoff, are sometimes sensitive to the method of the registration of denaturation changes (changes in solubility, absorbtion spectrum, viscosity, optical activity, enzymatic activity, stability in relation to proteolysis, availability of certain chemical groups), to the type of denaturation agent (heat, acids, urea, etc.), to minor changes in ion concentrations. Corresponding examples can be found in many reviews, concerning the thermodynamic aspects of denaturation (see, e.g. [5.9]).

The same situation is quite common as regards activation parameters: activation energy ΔE_a and activation entropy ΔS_a. Let us discuss several examples, illustrating the measurements of denaturation thermodynamic parameters by means of equilibrium and thermochemic (direct) methods, and the measurements of activation parameters by means of kinetic methods.

5.2.1 Acid Denaturation of Ferrihemoglobin

Direct measurements of the heat effect of this reaction were carried out at 15 and 25°C [5.10]. At 15°C, within the pH range of 3.51-3.19, the rate constant increases from 0.031 to 0.130 min^{-1}, and $\Delta H = -76 \pm 1.6$ kcal/mol^{-1}. At 25°C, within the pH range of 3.79-3.40, the rate constant increases from 0.029 to 0.350 min^{-1}, the heat effect changes its sign and becomes $+10.0 \pm 0.3$ kcal/mol^{-1} (the positive sign corresponds to heat absorption). Denaturation is accompanied by proton binding to the same extent at both temperatures, and, therefore, one must take into account the energy of proton transfer from the medium (+12.5 kcal/mol^{-1} at 15°C and 0 at 25°C). The difference between reaction heats for 15° and 25° becomes then even greater

$$\Delta H^{15} = -88.5 \text{ kcal/mol}^{-1} \ , \quad \Delta H^{25} = +10.0 \text{ kcal/mol}^{-1} \ . \tag{5.1}$$

Thus, with an increase in temperature by 10°, the reaction heat increases almost by 100 kcal/mol and changes its sign. It formally corresponds to an increase in specific heat by $\Delta c_p = +9850$ cal/mol^{-1} deg^{-1} (at 20°C). Almost the same values were obtained for serum albumine [5.11].

What can be the cause of such strong temperature dependence of the heat effect? It can be explained in two different ways. In the first place, this result may be totally determined by the difference between the specific heats of the native and the denaturated state (Kirschhoff law). In this case, the value has a real physical meaning and cannot be regarded as formal. This difference in specific heats may be, for instance, ensured by the expelling of hydrophobic groups on to the globule surface in the course of denaturation, and by a consequent increase of hydrophobic interaction. The alternative explanation suggests that at different temperatures we observe essentially different processes of acid denaturation. It means that either the initial or the final structure (or both) undergo temperature changes accompanied with heat absorbtion or generation. According to this view, different acid denaturation reactions with different heat effects are realized at different temperatures. What is the principal distinction between these possible explanations? In the first case, the change of reaction heat with temperature, $(\partial \Delta H / \partial T)$, is determined by the difference in specific heats between the initial and the final state, (Δc_p). In our example

$$\frac{\partial \Delta H}{\partial T} = \Delta c_p = 9850 \text{ cal/mol}^{-1} \text{ deg}^{-1} \ . \tag{5.2}$$

All structural changes in the system occur exclusively as a result of the denaturation process (in this kind of experiment, due to the pH change at a fixed temperature value, and in another possible kind of experiment, due to the temperature change at fixed pH value). The temperature change does not in itself lead to any changes in the characteristics of the initial and the final states. In this case, an evident thermodynamic relation

$$\frac{\partial \Delta H}{\partial T} = T \frac{\partial \Delta S}{\partial T} \tag{5.3}$$

connecting the temperature coefficients of the reaction heat effect and entropy change must be valid.

If the second explanation is true, then, at different temperatures, different reactions are proceeding, and (5.3) will not hold true. In this case, ΔH and ΔS must include the quantities $\Delta H'$ and $\Delta S'$, corresponding to the temperature-induced structural changes of the initial (or final, or both) compounds. Quantities $\Delta H'$ and $\Delta S'$ are not interconnected by (5.3). Let, for example, the process of acid denaturation at T_1 be accompanied by enthalpy change $\Delta H(T_1)$ and entropy change $\Delta S(T_1)$. Let us now carry out the same process at $T_2 > T_1$. Certainly, ΔH and ΔS

will change according to Kirschhoff law depending on value. If, however, the structure of the initial compound has been changed, i.e., if due to heating, this compound has undergone a transformation with thermodynamic parameters $\Delta H'$ and $\Delta S'$, the observed values of thermodynamic parameters of acid denaturation become equal to $\Delta H(T_2) - \Delta H'$ and $\Delta S(T_2) - \Delta S$ (for the sake of simplification, we assume that the structure of the final denatured state does not change with temperature). Since $\Delta H'$ and $\Delta S'$ are determined by specific protein structure and do not obey (5.3), the observed acid denaturaion parameters $\Delta H - \Delta H'$ and $\Delta S - \Delta S'$ will not obey (5.3) either. If this second explanation is true, the structure of the protein we are dealing with must, in the room temperature range, undergo changes affecting essentially the reaction path of acid denaturation.

Investigation of the same process of acid denaturation by means of equilibrium methods has produced quite different results. STEINHARDT and ZAISER [5.12,13] measured spectrophotometrically the equilibrium constant K of denaturation (denaturation leads to changes in the visual absorption spectrum of ferrihermoglobin), and, using Van't.Hoff equations

$$K = \exp\left(\frac{\Delta S}{R}\right)\exp\left(-\frac{\Delta H}{RT}\right) \tag{5.4}$$

$$\frac{d \ln K}{d(1/T)} = -\frac{\Delta H}{R} \tag{5.5}$$

found that the plot $\ln K = f(1/T)$ had two linear parts with one break at $15.5^\circ C$. In the temperature range of $0.2 - 15.5^\circ C$, reaction heat is $+32$ kcal/mol^{-1}, and between 15.5 and $25^\circ C$ it is essentially zero.

This great discrepancy between the calorimetric and the equilibrium measurements seems to be rather strange, if we assume that the heat effect temperature dependence. is determined exclusively by the difference between specific heats of the initial and the final state, i.e., by Kirschhoff law. It is sometimes suggested that determining the reaction heat by the equilibrium constant temperature dependence (i.e., from the slope of the curve in coordinates $\ln K$ and $1/T$) requires temperature independence of ΔH and ΔS. It is, however, not so. Indeed, let ΔH and ΔS be not dependent on temperature. Then, from (5.4) we have

$$\frac{dK}{dT} = K\frac{\Delta H}{RT^2} \tag{5.6}$$

is the equivalent of (5.5).

If ΔS and ΔH are temperature dependent,

$$\frac{dK}{dT} = K\left(\frac{\Delta H}{RT^2} - \frac{1}{RT}\frac{\partial \Delta H}{\partial T} + \frac{1}{R}\frac{\partial \Delta S}{\partial T}\right) \ . \tag{5.7}$$

Fulfillment of thermodynamic relation (5.3) turns (5.7) into (5.6). Thus, the foundations of (5.5) do not involve the requirement that enthalpy and entropy changes, in the course of the process investigated, be independent of temperature but do involve the requirement of (5.3) being valid. In particular, for Kirschhoff's case

$$\Delta H = \Delta H^{\circ} + \Delta c_p T \quad , \quad \Delta S = \Delta S^{\circ} + \Delta c_p \ln T \quad . \tag{5.8}$$

When the temperature dependence of reaction enthalpy and entropy is determined by the difference in heat capacities for the final and initial states only, the Van't Hoff equation should give true ΔH values.

The observed great discrepancies between the results of calorimetric and equilibrium measurements, differing even by their sign, allow no easy explanation.

FOREST and STURTEVANT [5.10] tried to interpret these discrepancies by assuming that ferrihemoglobin acid denaturation proceeds in two stages, one of which cannot be recorded spectrophotometrically. Let K_1 and K_2 be the equilibrium constants of these two stages, and ΔH_1 and ΔH_2 the corresponding true (i.e., measured calorimetrically) molar reaction heats

$$\begin{array}{cc} K_1 & K_2 \\ A \rightleftarrows B \rightleftarrows C & . \\ \Delta H_1 & \Delta H_2 \end{array} \tag{5.9}$$

Let us assume now that only the second transition can be recorded spectrophotometrically, and the spectra of A and B species are indistinguishable (the researcher does not even suspect that there is an intermediate state B). Let us call this case "situation 1".

Let b_e and c_e be the true equilibrium concentrations of B and C, respectively, and a_0 the total protein concentration. Then, in the case of situation 1, spectrophotometric measurements give

$$K_{app}^{(1)} = \frac{c_e}{a_0 - c_e} \quad . \tag{5.10}$$

Evidently

$$\frac{b_e}{a_0 - b_e - c_e} = K_1 \quad , \quad \frac{c_e}{b_e} = K_2 \tag{5.11}$$

and

$$K_{app}^{(1)} = \frac{K_1 K_2}{1 + K_1} \quad . \tag{5.12}$$

Using the Van't Hoff equation

$$\frac{dK}{dT} = K \frac{\Delta H}{RT^2} \tag{5.6}$$

we easily obtain

$$\Delta H_{app}^{(1)} = \Delta H_2 + \frac{\Delta H_1}{1 + K_1} \quad . \tag{5.13}$$

Let us designate as "situation 2" another possible case when only the first transition is recorded, and the spectra of B and C are indistinguishable[1]. Then

$$K_{app}^{(2)} = \frac{b_e + c_e}{a_0 - b_e - c_e} = K_1 + K_1 K_2 \tag{5.14}$$

and

$$\Delta H_{app}^{(2)} = \Delta H_1 + \frac{K_2}{1 + K_2} \Delta H_2 \quad . \tag{5.15}$$

Calculate now the true denaturation enthalpy value obtained by calorimetric measurements. Assume that initially there was 1 mol of form A, and at the end of the process there appeared equilibrium quantities b_e of form B and c_e of form C. It means that $b_e + c_e$ moles of form A have undergone transformation 1 (with molar reaction heat ΔH_1), and then the c_e moles of B have undergone transformation 2 (with molar reaction heat H_2). Then

$$\Delta H_{true} = (b_e + c_e)\Delta H_1 + c_e \Delta H_2 \quad .$$

Taking (5.11) into account and assuming $a_0 = 1$ we have

$$b_e = \frac{K_1}{1 + K_1 + K_1 K_2} \quad , \quad c_e = \frac{K_1 K_2}{1 + K_1 + K_1 K_2}$$

and

$$\Delta H_{true} = \frac{K_1(1 + K_1) H_1 + K_1 K_2 \Delta H_2}{1 + K_1 + K_1 K_2} \quad . \tag{5.16}$$

If the process proceeds almost to completion, i.e., if $K_1, K_2 \gg 1$

$$\Delta H_{true} = \Delta H_1 + \Delta H_2 \tag{5.17}$$

and

$$\Delta H_{app}^{(1)} = \Delta H_2 \quad , \quad \Delta H_{app}^{(2)} = \Delta H_1 + \Delta H_2 \quad . \tag{5.18}$$

Thus, if the equilibrium position is shifted far to the right, we obtain then in the case of situation 1, using the Van't Hoff equation, the heat of the second stage, and, in the case of situation 2, we must obtain the true enthalpy value. Determinations of equilibrium constants for Van't Hoff enthalpy calculations are carried out with different relative concentrations of native and denatured proteins (chang-

---

[1]In [5.10] FOREST and STURTEVANT considered an unrealistic situation when the spectra of A and C are indistinguishable. For the acid denaturation of ferrihemoglobin (as for all other cases) it cannot be true: in this case, the experimentator would be unable to record the complete denaturation.

ing the degree of denaturation by means of pH or temperature), including systems
with the denaturation equilibrium shifted far to the right. Therefore, the above-
mentioned sharp discrepancies between the results of calorimetric and those of
equilibrium measurements permit us to disregard situation 2. Indeed, two enthalpy
values are not only quite different but shift with temperature in different directions. In the case of situation 1 when the first stage in (5.9) is not recorded,
with the equilibrium shifted to the right, the apparent enthalpy of the process
should be equal, according to (5.18), to the true enthalpy of the second stage.

Thus, in this case

$$\Delta H_1^{25} = +10 \text{ kcal/mol}^{-1} \quad \Delta H_2^{25} = 0 \tag{5.19a}$$

$$\Delta H_1^{15} = -120.5 \text{ kcal/mol}^{-1} \quad \Delta H_2^{15} = +32 \text{ kcal/mol}^{-1} \quad . \tag{5.19b}$$

If the temperature dependence of K_{app} is measured when $K_{app} \approx 1$ (i.e., at pH values
such that denaturated and native form concentrations are comparable), it is easy
to show that the experimental data for ΔH_{app} and ΔH_{true} ($\Delta H_{true} = \Delta H_1 + \Delta H_2$) can
be consistent only if

$$\Delta H_1^{25} = 10(1 + 1/K_1)\text{kcal/mol}^{-1}$$

$$\Delta H_2^{25} = -10/K_1 \text{ kcal/mol}^{-1} \tag{5.20a}$$

$$\Delta H_1^{15} = -120.5 (1 + 1/K_1)\text{kcal/mol}^{-1}$$

$$\Delta H_2^{15} = 32 + 120.5/K_1 \text{ kcal/mol}^{-1} \quad . \tag{5.20b}$$

Equations (5.19,20) are compatible if, at both temperatures, $K_1 \to \infty$, i.e., if
the equilibrium A$\rightleftarrows$B in (5.9) is completely shifted to the right. This is quite
improbable because at 15 and 25° the heats of this process must have different signs
and must differ by $\sim$130 kcal/mol^{-1} [see (5.19)]. The representation of ferrihemoglobin acid denaturation as a two-stage reaction cannot, thus, explain the discrepancies between the calorimetric and the equilibrium measurements. All this casts
doubt on the possibility of applying the Van't Hoff equation in the cases of protein
denaturation.

The most important result of the investigations discussed above is, probably,
the proof of the fact that temperature changes, in the physiological range of
temperatures, lead to considerable changes in the protein molecular structure, accompanied by its enthalpy changes. It follows from the sharp changes of the calorimetrically measured reaction heat in the narrow temperature range between 15 and
25°C.

Direct experimental evidence of structural changes in globular proteins with
temperature in the physiological range was provided by the calorimetric investigations of protein thermal denaturation processes.

5.2.2 Thermal Denaturation of Proteins

Among many kinds of denaturation, thermal denaturation is unique in the sense that, in this case, to determine the thermodynamic parameters, one measures the dependence of the degree of transformation on temperature, i.e., on the factor whose changes lead to the transformation in question. This must be borne in mind when thermodynamic aspects of protein thermal denaturation are discussed. The most essential results were, probably, obtained using the method of microcalorimetry and comparing these data with those of equilibrium measurements. Due to the importance of this problem, we shall discuss in detail two works on the thermodynamics of globular protein thermal denaturation, carried out in two leading laboratories. Let us begin with the results obtained by JACKSON and BRANDTS [5.14] concerning the reversible thermal denaturation of chymotrypsinogen. In a calorimetric study one measures the temperature dependence of the partial heat capacity c_p of a protein. A typical differential calorimetric curve is shown in Fig.5.1a. The integral form of the same curve is shown in Fig.5.1b; the solid line in Fig.5.1b shows the temperature dependence of protein partial enthalpy, H, in solution. For native chymotrypsinogen at 20°C the c_p(nat) value is equal to 0.40 cal/g^{-1} deg^{-1} and has a strong temperature dependence: temperature increase by 1° results in the increase of c_p by 0.7%.

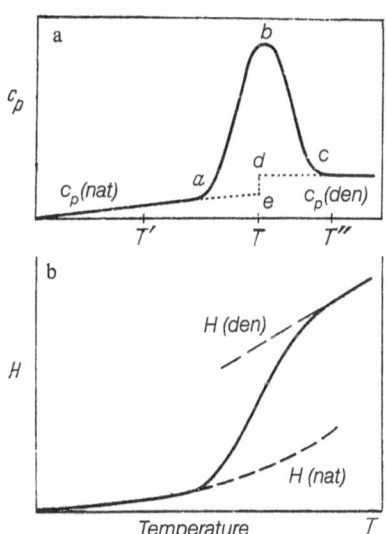

Fig.5.1 a,b. Typical differential (a) and integral (b) calorimetric curves for chymotrypsinogen in the region of thermal denaturation (according to [5.2])

This temperature dependence of heat capacity is preserved in dry chymotrypsinogen samples and, therefore, cannot be explained by the changes in protein-solvent interaction but is, essentially, a protein property. For waterless chymotrpysinogen the heat capacity value is by 30% lower than that for chymotrypsinogen in solution, but the temperature coefficient is practically the same (0.6% instead of 0.7%). When the temperature approaches that of denaturation, the increase of C_p with temperature accelerates, heat capacity passes through a maximum, and then drops almost to its initial value.

The temperature coefficient of denatured chymotrypsinogen c_p(den) heat capacity is considerably less than for the native protein. The enthalpy change in the course of denaturation is calculated as area abcdea in Fig.5.1a

$$\Delta H_{cal}^{den}(T) = \int_{T'}^{T''} c_p dT - \int_{T'}^{T} c_p(nat)dT - \int_{T}^{T''} c_p(den)dT \quad . \tag{5.21}$$

To perform this procedure, it is necessary to extrapolate c_p(nat) and c_p(den) into the transition region, and a denaturational heat capacity jump de appears. Such a procedure is not unique. We shall see later that other authors have obtained $\Delta c_p = 0$ using c_p(nat) extrapolation for the same protein. Subdividing the whole temperature interval into two regions is equivalent to defining the denaturation process as a transition between two states — the native and the denaturated. The native state exists at all temperatures lower than that of the beginning of the "main" denaturation transition (point a in Fig.5.1a), being by definition a region of great changes in various physical properties, and the denaturated state exists at all temperatures higher than that of the end of this transition. Temperature variation of c_p within the regions corresponding to the native and the denaturated states is interpreted as a state characteristic (and not as its change) independent of the nature of this variation: is it due either to trivial thermal excitations of the internal degrees of freedom of a macromolecule, or to conformational transitions, accompanied by heat absorbtion or by changes in protein-solvent interaction (see Sect.5.6). With this definition of the initial and the final macromolecular states, between which a denaturation transition takes place, the enthalpy values of this transition, measured calorimetrically (as abcdea area in Fig.5.1a) will not differ appreciably from Van't Hoff equilibrium values ΔH_{VH}^{den}; these are to be estimated using the same plot of $K = f(T)$ [e.g. (5.6)], equilibrium constant K being the ratio of corresponding areas under the curve, in Fig.5.1a. ΔH_{cal}^{den} and ΔH_{VH}^{den} values obtained for chynotrypsinogen at different pH are listed in Table 5.1.

Table 5.1

pH	T[°C]	ΔH_{cal}^{den} [kcal/mol^{-1}]	Δc_p [cal/mol^{-1} deg^{-1}]	ΔH_{VH}^{den} [kcal/mol^{-1}]
1.95	40,6	103	4200	94
2,03	42,0	102	3300	98
2,08	42,0	99	3900	98
2,59	48,0	126	3200	126
2,99	53,9	145	2800	133
3,02	54,2	135	2800	134

The calculated equilibrium constant values do not lie, in Van't Hoff's coordinates ($\ln K$ and $1/T$), on one straight line (it is due to Δc_p being unequal to zero). ΔH_{VH}^{den} values listed in Table 5.1 correspond to the temperatures of c_p maxima (Fig.5.1a).

As can be seen from Table 5.1, there is quite a good coincidence between the calorimetric and the equilibrium values. The Δc_p jump at transition temperature is explained by the protein conformational change (the expulsion of hydrophobic groups to the globule surface) in the course of denaturation, but the no less essential changes of $c_p(nat)$ in the low temperature range are not considered to be due to a change in the molecular state.

This coincidence between ΔH_{cal}^{den} and ΔH_{VH}^{den} does, according to the authors, suggest the one-step nature of the denaturation process.

The most interesting result of this study is, probably, the demonstration of the existence of essential heat capacity changes for native protein in the physiological temperature range, much lower than the temperature interval of the main denaturation transition. These c_p changes (recorded, by the way, by many authors for all the proteins studied) are observed within broad temperature intervals and are rather smooth: abrupt predenaturational changes of c_p were never observed.

The thermal denaturation of chymotrypsinogen (and of certain other proteins) was also studied calorimetrically by PRIVALOV et al. [5.15]. The technique applied and the experimental curves obtained did not, practically, differ from those described above. The treatment of these data was, however, carried out differently. Figure 5.2 shows typical calorimetric curves for various pH values. As in the previously cited work, chymotrypsinogen heat capacity at 20°C equals 0.4 cal/g^{-1} deg^{-1}. But, according to PRIVALOV and his co-workers, $c_p(nat)$ does not change up to 28°C and after that increases for all pH values. The authors related the observed heat capacity changes to the predenaturational and denaturational stages, respectively. When protein changes are registered by means of extinction measurements at 2820A, the same two stages can be observed (Fig.5.3). The precise division into two stages is carried out in such a way that a straight line (in Van't Hoff's coordinates) is to give a good fit to experimental spectrophotometric data. The temperature independence of ΔH_{VH}^{den} is, thus, postulated (contrary to [5.14]). The authors also postulated that heat capacity changes occur only during the first stage (dotted line in Fig.5.2), and that, therefore, a denaturational heat capacity jump does not occur in this case. With this assumption, the calculated values of denaturation enthalpy ΔH_{cal}^{den} are quite close to the equilibrium ones (Table 5.2).

One can accept another assumption concerning the true heat capacity temperature dependence; c_p, for instance, can change only during the second, denaturational, stage of the process, or during both stages in direct proportion to the absorbed heat. In the latter case, total enthalpy values (see Table 5.2) are obtained. By subtracting the corresponding ΔH_{VH}^{den} value form ΔH_{cal}^0 one can estimate the heat ab-

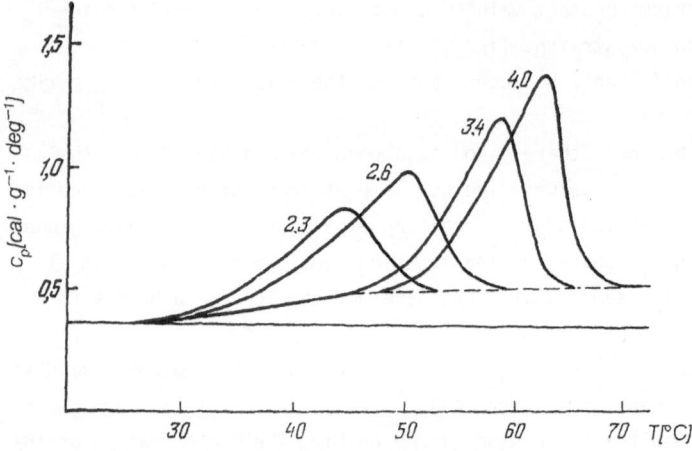

Fig.5.2. Calorimetric curves for chymotrypsinogen at various pH values (according to [5.18]). Numbers at curves give the pH values

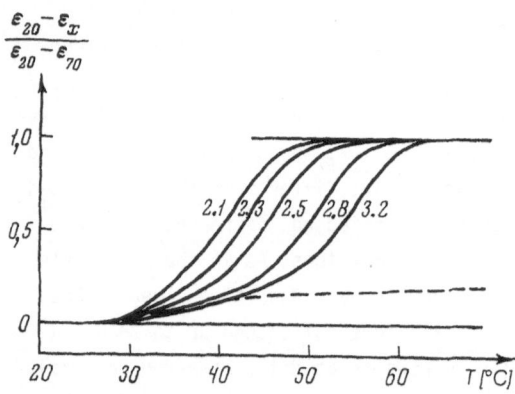

Fig.5.3. Temperature dependence of extinction at 2820 Å for a chymotrypsinogen solution at various pH values

Table 5.2

pH	ΔH_{vH}^{den} [kcal/mol^{-1}]	ΔH_{cal}^{den} [kcal/mol^{-1}]	ΔH_{cal}^{0} [kcal/mol^{-1}]
2,3	75	78	105
2,6	100	102	140
2,8	105	110	155
3,4	132	130	182
4,0	143	140	200
5,0	145	148	207

sorbed during the first, predenaturational, stage (30-60 kcal/mol^{-1}). The values
obtained do not exactly coincide with the first stage heat effect estimates, based
on transition sharpness according to the Van't Hoff equation. The reliability of
these estimations is, of course, rather poor: the first and the second stages are
partially overlapping, and the separation of the denaturation stage by its enthalpy
temperature independence is rather unfounded. Nevertheless, the authors accepted
that c_p changes occur only during the predenaturation stage. Due to the partial
overlapping of this stage with the next one, it was impossible to estimate the
area corresponding to the heat absorbed. Therefore, the authors have made a rather
voluntary conclusion that during the predenaturational stage heat is neither absorbed nor evolved, and only the system heat capacity changes. It is doubtful,
however, that such strong heat capacity temperature dependence can be caused by the
thermal excitation of intramolecular degrees of freedom only. The authors themselves
assumed that heat capacity changes are caused, mainly, by the changes in the number
of contacts between hydrophobic protein group and solvent molecules (incidentally,
JACKSON and BRANDTS [5.14], used the same reasoning to explain the c_p denaturation
jump postulated in their work). Considering the experimental data with an open mind
one can picture the following pattern of the whole process. With increasing temperature of solution protein enthalpy changes begin already at room temperature and are
considerably greater than could be expected if they were due to the trivial heat
capacity increase in consequence of the thermal excitation of intramolecular degrees of freedom. In a certain temperature region the enthalpy changes become more
intensive and then slow down again. The changes of system enthalpy at the predenaturational stage are comparable with those of denaturation.

One way or another, both types of changes are accompanied by changes in the
macromolecule structure (which influence the interaction between the macromolecule
and the solvent). The predenaturation stage may correspond to configurational
changes, and the denaturation stage to conformational changes (see Sect.4.4).

Recently PRIVALOV [5.16-19] has again described calorimetric measurements of
predenaturational changes and of thermal denaturation for several globular proteins
(lysozyme, α-chymotrypsin, α-chymotrypsinogen, myoglobin, ribonuclease and cytochrome c). In comparison with [5.15], there are certain rather insignificant changes
in experimental data and their interpretation. In the midpoint of denaturation there
is now a heat capacity jump as that described in [5.14]. The linear increase of c_p
with temperature is now observed beginning from the lowest temperatures studied
(20°C). According to these papers, protein changes during heating proceed in two
qualitatively different stages: the stage of the linear increase of heat capacity,
and the stage of intensive heat absorption. The enthalpy change during the second,
denaturational, stage (as always, the division of one curve into parts corresponding
to different stages introduces some uncertainty in the quantitative estimations),
practically coincides with the equilibrium value calculated from the calorimetric
curve by means of the Van't Hoff equation.

According to the important conclusion of PRIVALOV, the predenaturational stage cannot be regarded as a trivial heat capacity temperature change. Absorption spectra of cytochrome C, in predenaturational and denaturational temperature intervals, were carefully measured, and the appearance of quite distinct isobestic points in the predenaturational region (20-65°C at pH 3.0) was especially emphasized. The obvious conclusion is that two interconversionable conformational states coexist within this temperature interval. The possibility of several consecutive conformational transitions cannot, however, be excluded. Indeed, not every conformational transition must necessarily lead to the change in the heme group absorption spectrum.

At first sight, it seems that there is a major discrepancy between the above data on the acid and the thermal protein denaturation. We have seen that in the case of ferrihemoglobin acid denaturation, the calorimetric and the equilibrium (Van't Hoff) heat effects are essentially different. At the same time, in the case of thermal denaturation, we are able to obtain quite a satisfactory agreement between calorimetric and equilibrium enthalpy values by means of a rather simple fitting process (a more or less arbitrary subdivision into separate stages or an arbitrary extrapolation of c_p(nat) changes to the region of denaturational transition). This discrepancy between the results of the two types of experiments is, however, quite understandable. We have seen that, in the case of acid denaturation, thermodynamic condition (5.3) does not hold true, and the Van't Hoff equation is inapplicable. The violation of (5.3) is due to the gradual structural changes of the macromolecule with temperature. These changes in macromolecular structure are accompanied by enthalpy changes, and may be regarded as a chemical transformation with specific ΔH and ΔS values not interconnected by (5.3). Therefore, when acid denaturation at different temperatures is studied, the resulting temperature dependence of the measured heat effect will be determined not only by the heat capacity difference between the native and the denatured protein, but also by the changes of the native protein enthalpy content with temperature (and even by the possible changes in the path of the denaturation process) caused by the changes in native protein structure with temperature. As a result, the temperature dependence of the acid denaturation summary heat effect will not obey (5.3), the Van't Hoff equation will not hold true, and equilibrium measurements of ΔH and ΔS become meaningless.

We shall further see to what errors can, in this case, lead the introduction of experimental data into the Van't Hoff equation. At the same time, if the temperature is used not only for the drawing of the Van't Hoff plot, but serves as a parameter whose change initiates the process in question, the situation becomes entirely different. In this case (heat denaturation), the temperature change causes the studied transformation, and the initial conditions are chosen to be the same for the calorimetric and the equilibrium measurements. The determination of parameter interval values, that must correspond to the measured transition, is made, essentially, in such a way that (5.3) has to be fulfilled. It is, therefore, always possible to single out the part of overall transition, for which calorimetric and equilibrium measurements coincide (if the transition proceeds symmetrically in relation to temperature). When all temperature changes in initial and final states, that do not allow comparison of the calorimetric and the equilibrium data, are by one way or another assigned to intrinsic characteristics of these macroscopic states, it would be surprising if we *could not* single out that part of the transition which makes this comparison possible (see also Sect.5.6).

The main result of calorimetric investigations of globular protein thermal denaturation is, from my point of view, the demonstration of gradual predenaturational changes in a broad range of physiological temperatures. It must be mentioned that, for all the proteins studied so far, the calorimetric data reflect the monotony of these thermodynamic parameters — throughout a broad temperature interval.

Many experiments performed with the help of different methods may lead to the conclusion that quite a significant structural rearrangement of protein macromolecules can take place at comparatively low temperatures. Corresponding data and references can be found, for instance, in well-known monographs [5.20,21]. The greater part of examples concerns the abrupt changes in activation parameters of enzyme reactions within narrow temperature intervals (so-called "breaks" on the Arrhenius anamorphosis). These results will be discussed later. The data on the changes of other physical characteristics (degree of fluorescence polarization, rate constant of photochemiluminescence decay, photoinactivation quantum yield, position of maximum in fluorescence spectrum) for various proteins at temperature $10-40^{\circ}C$ can be found in [5.20]. The slopes of curves in the corresponding transition regions are usually not very steep, and it is very difficult to maintain that there is a strict coincidence of transition temperatures for one and the same protein studied with the help of different methods. One cannot, however, doubt that some structural rearrangement of protein macromolecules in the investigated temperature region does actually take place.

Calorimetric studies of protein denaturation processes [5.22-24] have also shown a strong temperature dependence of true heat effects, that can be interpreted as indicative of essential thermal rearrangements of these molecules outside the denaturation region too.

5.2.3 On the Kinetics of Denaturation Processes

Measurements of kinetic parameters of protein denaturation were conducted by many authors. Almost all of them treat the kinetic results (as it is always the case in biochemical kinetics) by means of the Arrhenius equation

$$\kappa = A \exp(-\Delta E_a/RT) \tag{5.22}$$

and by applying the activated complex theory (the Polany-Eyring theory)

$$\kappa = (kT/h)\exp(\Delta S_a/R)\exp(-\Delta E_a/RT) \tag{5.23}$$

where κ is the reaction rate constant, ΔE_a-activation energy or enthalpy[2], ΔS_a - activation entropy, k - Boltzmann constant, h - Planck constant, and R - gas constant.

Assuming that the rate constant temperature dependence is determined exclusively by the exponential factor in (5.22) one can find ΔE_a and A, and then calculate ΔS_a from (5.23).

Thus, the basic parameters, whose values help in usual meditations on the mechanism and kinetic properties of a denaturation process, appear to be ΔE_a and ΔS_a. If large positive ΔS_a values have been obtained, an author usually speaks about system "randomization" in the activated state, evaluates the relative significance of energy and entropy factors in the process, etc.

One must understand that (5.22,23) are based on the assumption of the existence of an activated state in thermodynamic equilibrium with the initial state. E_a must have the meaning of heat effect for the reaction of system transition from initial to activated state. We have already seen that heat effects of denaturation processes, determined by equilibrium measurements using the Van't Hoff equation [which is completely identical with the Arrhenius equation (5.22)], may have nothing in common with their true values. In the case of thermodynamic parameters, however, it is always possible to measure these heat effects calorimetrically. For kinetic activation parameters there is no such possibility, and we must be extremely cautious with the values and the physical meaning of these parameters obtained with the help of the Arrhenius equation. Existing literary data justify this suspicion. Let us consider only one example.

OKUNUKI [5.25] studied the denaturation kinetics of crystalline α-amylase from *Bacillus subtilis* with different denaturating agents. The degree of denaturation was determined by the rate of protein enzymatic hydrolysis under the action of proteinase isolated from the same culture (denaturation leads to hydrolysis acceleration). Denaturation rate curves could be satisfactorily represented by first-order kinetic equations. The temperature dependence of rate constants did obey the Arrhenius law. The results are given in Table 5.3 (k and F_a values are given for 50°C, pH was 8.0, and in the case of acid denaturation, 4.5).

It can be seen that, for different methods of denaturation, the actually observed rate of process is changing not more than by 3 orders of magnitude, and the F_a value unequivocally connected with rate constant values, not more than by 4.7 kcal/mol^{-1}.

---

[2]Some authors make a distinction between energy, ΔE_a, and enthalpy, ΔH_a, of activation, and assume that $\Delta H_a = \Delta E_a - RT$. In this case, ΔS_a in (5.23) increases by R. It can be shown that this distinction is the result of a misunderstanding (see Sect.5.3).

Table 5.3

Denaturating agent	k $[s^{-1}]$	ΔF_a $[kcal/mol^{-1}]$	ΔH_a $[kcal/mol^{-1}]$	ΔS_a $[eu]$
6M urea	$1,73 \times 10^{-4}$	24,6	32,8	25,4
6M urea + 0.05M $CaCl_2$	$3,95 \times 10^{-5}$	25,5	32,8	22,4
Heat	$1,12 \times 10^{-5}$	26,4	67,6	127,5
Heat + 0.05M $CaCl_2$	$4,18 \times 10^{-7}$	28,4	67,6	121,5
Acid	$7,07 \times 10^{-4}$	23,7	39,5	49,5
Acid + 0.05M $CaCl_2$	$5,10 \times 10^{-5}$	25,4	39,5	43,7

At the same time, the values of ΔH_a and ΔS_a, calculated with the help of (5.22,23), change by 34.8 kcal/mol^{-1} and 105.1 eu, respectively. It means that exponential and preexponential factors in (5.22) both change by 24 orders of magnitude. The enourmous magnitude of this change, that cannot be verified by any direct method of measurement, seems to be rather suspicious.

The rates of conformational changes, accompanying protein renaturation after the action of the denaturing agent has been abruptly stopped, were measured by several authors. These intramolecular transformations proved to be comparatively slow. For instance, conformational changes of acid denaturated staphylococcal nuclease, after the abrupt rise of pH to a neutral value, were recorded by the changes in triptophane fluorescence [5.26]. In this case, conformational changes proceed in two stages: a temperature-dependent fast stage with $\tau_{\frac{1}{2}} \approx 0.055$ s, and a temperature-independent slow stage ($\tau_{\frac{1}{2}}$ diminishes from 0.60 to 0.15 s when temperature is raised from 13 to 38°C). Apparently, due to the first stage, the protein does not return to its initial state, but acquires another, kinetically easier available state, slowly relaxing after that to the initial one.

TEIPEL and KOSHLAND [5.27,28] studied the renaturation of various enzymes with the help of optical methods as well as by enzymatic activity restoration. They found even slower processes (the first stage, accompanied by main structural changes, actually lasted ~1 min, and the second stage, during which the activity was restored but the structural changes were comparatively small, required about 100 min).

5.3 On the Difference Between Activation Energy and Activation Enthalpy

In many monographs and reviews dealing with transition-state theory of chemical reaction rates (see, e.g. [5.29,30]) there appears the formula

$$\kappa = e \frac{kT}{h} \exp\left(\frac{\Delta S_a}{R}\right) \exp\left(-\frac{E_{expt}}{RT}\right) \tag{5.24}$$

into which the basic equation of transition-state theory[3]

$$\tilde{\kappa} = \frac{kT}{h} \exp\left(\frac{\Delta S_a}{R}\right) \exp\left(-\frac{\Delta H_a}{RT}\right) \tag{5.25}$$

is transformed if "the experimental activation energy", ΔE_{expt}, is used in place of theoretical "enthalpy of activation", ΔH_a.

Let us repeat the derivation of (5.24) that can be found in [5.29] and other works. The experimental activation energy is determined from the temperature dependence of the reaction rate constant by means of the Arrhenius empirical equation

$$\kappa = A \exp\left(-\frac{\Delta E_{expt}}{RT}\right) \tag{5.26}$$

$$\frac{d \ln \kappa}{dT} = \frac{\Delta E_{expt}}{RT^2} . \tag{5.27}$$

On the other hand, from (5.25)

$$\frac{d \ln \kappa}{dT} = \frac{1}{T} + \frac{\Delta H_a}{RT^2} . \tag{5.28}$$

Comparing (5.27) and (5.28) we obtain

$$\Delta H_a = \Delta E_{expt} - RT . \tag{5.29}$$

Substituting (5.29) into (5.25) we obtain (5.24).

Let us try to elucidate the meaning of the different steps of this derivation. The entropy and the enthalpy of activation in theoretical equation (5.25) are constants[4] whose values for a given system can be (in principle) calculated. In empirical equation (5.26) A and ΔE_{expt} are constant by definition.

It is clear that the curves described by (5.25,26) cannot coincide at any values of the constants ΔH_a, ΔS_a, A and ΔE_{expt}. The equalization of (5.27,28) has only the following obvious meaning. For given values of two mutually independent parameters ΔE_{expt} and ΔH_a a temperature can be found for which the slopes of functions (5.24,26) in the coordinates $\ln \kappa - 1/T$ are equal.

Thus, (5.29) is the equation to be used for the estimation of the temperature point at which the logarithmic plots of curves (5.25,26) have the same slopes. There is no physical meaning in substituting (5.29) into (5.25) with "T" as a variable.

---

[3]It is sometimes stated that (5.24) is valid only for monomolecular gas reactions and for all the reactions in condensed phase for which activated complex formation is not accompanied by changes in specific volume, otherwise, instead of factor e, factors e^2, e^3, etc., may appear. Here we shall restrict ourself to the simplest case, i.e. (5.24).

[4]Or they are temperature dependent but interconnected by (5.3). This would not introduce any principal changes into our analysis.

When an experimentalist applies the transition state theory (i.e., when he cal-
culates ΔS_a from his data) substituting ΔE_{expt} for ΔH_a in (5.25), he tacitly assumes that the activation barrier is equal to the enthalpy of transition from
initial products to the activated state, and that the temperature dependence of
the frequency factor is negligible, due to the narrowness of the temperature interval and to the experimental errors. There is no other meaning in the use of Arrhenius
activation energy in the formulas of transition state theory.

5.4 Some Protein Reactions

There are reliable calorimetric data concerning the complex formation of proteins
with low molecular compounds and other proteins which are also indicative of structural changes in the low temperature region. Here are several examples.

ADAMS and WEISS [5.31] carried out a calorimetric study of the reaction between
hemoglobin and haptoglobin — a globular protein (molecular weight 85000) that increases the peroxydase activity of hemoglobin. Complex formation enthalpies, ΔH,
at temperatures 4, 20 and 37°C were found to be +7.2, -29.7, and -70.2 kcal/mol^{-1},
respectively[5]. The observed effect is of the same order of magnitude as that for
protein denaturation, but has the opposite sign: transformation enthalpy becomes
positive at lower temperatures. The authors interpreted this result, in the same
way as in the case of denaturation, assuming large heat capacity differences between the initial and the final states due to the changing number of contacts between the hydrophobic groups and the solvent. In order to explain the opposite sign
of the observed effect, the authors were compelled to assume that the hydrophobic
groups in the associate are situated even deeper within the globule, i.e., that,
in this respect, the associate is more native than the original molecules.

As early as in 1966, CLARKE [5.33] measured the temperature dependence of this
reaction equilibrium constant using haptoglobin fluorescence quenching due to its
titration with hemoglobin. In the temperature range of 5-30°C the equilibrium
constant was, practically, temperature independent. Thus, the application of the
Van't Hoff equation for this reaction gives $\Delta H = 0 \pm 2$ kcal/mol^{-1}, i.e., a value
that has nothing in common with the true, calorimetrically measured, reaction
enthalpy.

There are many publications dealing with the calorimetry of inhibitor binding
to some globular enzymes. SHIAO and STURTEVANT [5.34] measured the heat effects
of indoll, N-acetyl-D-tryptophan and proflavin binding to α-chymotrypsin, and obtained, at pH 7.8 and 25°C, -25.2, -19.0, and -11.3 kcal/mol^{-1}, respectively. At

---

[5]Recently these measurements have been repeated [5.32]. The reaction appears to be
more complicated than it was assumed by ADAMS and WEISS, but the principal quantitative results are the same.

the same time, the equilibrium constant of N-acetyl-D-tryptophan reaction with α-chymotrypsin at pH 7.8 and 5°C [5.35] practically coincides with this equilibrium constant at pH 7.9 and 25°C [5.36], which means, according to the Van't Hoff equation, the zero enthalpy of binding.

YAPEL [5.37] measured the temperature dependence of equilibrium constants for reactions of indoll and N-acetyl-D-tryptophan with α-chymotrypsin and obtained, using the Van't Hoff equation, ΔH values of -22.6 and 9.8 kcal/mol^{-1}, respectively. Thus, equilibrium measurements performed by different authors and treated in accordance with the Van't Hoff equation, give, in this case, erroneous and contradicting results. SHIAO and STURTEVANT came to the conclusion that a considerable part of heat effects registered calorimetrically in the course of the binding of inhibitors with α-chymotrypsinogen is caused by protein conformational changes.

This conclusion was confirmed by systematic investigations of the thermodynamic parameters of the interaction between inhibitors and α-chymotrypsin carried out by SHIAO at varying conditions [5.38]. In all cases, inhibitor binding leads to protein conformational changes, these changes being the stronger the higher is the enzymatic activity (conformational changes at pH 7.8 are much more pronounced than at pH 5.6). Inhibitor binding is accompanied by changes in the number of protons attached to the protein, which can be regarded as additional evidence of conformational changes. Proton liberation is strongly temperature dependent and at 25°C is even transformed into proton absorption.

Similar effects were found by HINZ et al. [5.39] in the course of inhibitor (Hexitol-1,t-biphosphat) binding to aldolase. Let us discuss this work in detail. Inhibitor binding leads to an increase in the degree of protein protonization (the number of protons attached to one protein molecule increases at all temperatures studied by 1.4 ± 0.2) due to complex formation with the inhibitor at its saturation concentrations. The authors took account of the heat of proton extrication from the buffer, and determined the entropy of inhibitor binding, H_b, at its saturation concentrations, when each protein molecule binds 2.7 inhibitor molecules. Measurements did actually reveal a strong linear dependence (Table 5.4).

Table 5.4

T^0 [K]	278	283	288	293	298	303	308	313	318
ΔH_b [kcal/mol^{-1}]	23,3	17,8	12,3	6,8	1,3	-4,3	-9,8	-15,3	-20,8
ΔG_b^0 [kcal/mol^{-1}]	-5,91	-6,38	-6,75	-7,03	-7,23	-7,31	-7,32	-7,23	-7,05
ΔS_b^0 [eu]	104,8	85,2	65,9	47,0	28,4	10,1	-7,9	-25,6	-43,1
lg K_b	4,62	4,90	5,10	5,22	5,30	5,25	5,16	5,00	4,88

As is usually the case, the authors explained this dependence by protein heat capacity changes due to inhibitor binding and determined $\Delta c_p = -410$ cal deg^{-1} per mol of the bounded inhibitor. The following measurement and calculation procedure was then applied. The equilibrium constant of inhibitor binding by protein at 25°C (K_b) is determined by means of calorimetric titration

$$K_b = \frac{[EI]}{\{[E_0] - [EI]\}\{[I_0] - [E]\}} \quad . \tag{5.30}$$

Then, using the formula

$$\Delta G_b^0 = -RT \ln K_b \tag{5.31}$$

standard free energy of binding at 25°C is found, and, with the help of the Gibbs-Boltzmann equation,

$$\frac{\Delta G_b^{0(1)}}{T_1} - \frac{\Delta G_b^{0(2)}}{T_2} = \Delta H_b'\left(\frac{1}{T_1} - \frac{1}{T_2}\right) - \Delta c_p \ln \frac{T_1}{T_2} \tag{5.32}$$

where

$$\Delta H_b' = \Delta H_b - \Delta c_p T \approx 328.9 \text{ kcal/mol}^{-1} \quad , \tag{5.33}$$

ΔG_b^0 values for other temperatures are obtained. The values of ΔG_b^0, ΔS_b^0 and of K_b calculated with the help of (5.31) are also given in Table 5.4. After that the ΔS_b^0 values at various temperatures are calculated with the help of the conventional thermodynamical equation

$$\Delta S_b^0 = \frac{\Delta H_b - \Delta G_b^0}{T} \quad . \tag{5.34}$$

Let us now discuss the procedure and results of these calculations in greater detail. We are given to understand that direct experiments show the heat effect of low molecular inhibitor binding by protein to be strongly temperature dependent. It is assumed that the whole of this dependence is caused by the existence of a *constant* difference, Δc_p, between the heat capacities of the initial (2.7 molecules of free inhibitor + free protein) and the final (complex) states. Temperature independence of Δc_p is, thus, postulated, and all structural changes within the system are ascribed exclusively to inhibitor binding. The transition enthalpy value extrapolated to the absolute zero temperature ($\Delta H_{b'}$) is suspiciously high. A method of determining this value is illustrated in Fig.5.4 which shows the postulated temperature dependence of the enthalpies of initial (H_{free}) and final (H_{compl}) reaction products. It should be noted that the absolute slope of H(T) straight lines is arbitrary. Experiment, as well as calculations, determine Δc_p only, i.e., the angle between two straight lines.

Equation (5.32) is not actually necessary for the determination of ΔS_b^0 values. If the only cause of the temperature dependence of transition enthalpy is the

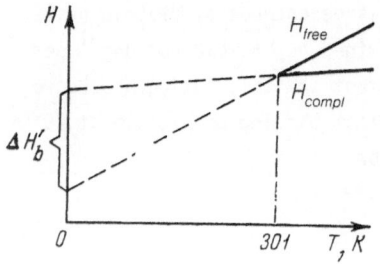

Fig.5.4.
Determination of $\Delta H_b'$ value

constant difference Δc_p, then, using (5.30,31,34) one can determine ΔS_b^0 value for 25°C (the temperature at which the equilibrium constant was actually measured), and by means of (5.8) find ΔS_b^0 for any other temperature

$$\Delta S_b^{0(1)} - \Delta S_b^{0(2)} = \Delta c_p \ln(T_1/T_2) \quad . \tag{5.35}$$

The results would, naturally, be the same. Thus, the treatment was based on two postulates: the fulfillment of (5.3), and Δc_p temperature independence. Taking into account all that has been previously said about structural changes in globular proteins at physiological temperatures, the validity of both postulates seems to be very improbable.

Using such calculations the authors obtained the values of equilibrium constants of complex formation at different temperatures (Table 5.4). Naturally, these values do not lie on a straight line in coordinates $\lg K_b$ and $1/T$. Moreover, K_b has a maximum at 25°C (i.e., at the only temperature, for which this quantity was actually measured). It would also be of interest to measure K_b directly at other temperatures.

This work was discussed in such detail only as an example. Similar results were obtained for other systems, as, for instance, in the case of complex formation between coenzyme nicotinamide-adenine dinucleotide and enzyme glyceraldehyde-3-phosphate dehydrogenase [5.40]. The data were treated in the same way as in the work discussed above (equilibrium measurements were made by registering the decay of protein tryptophan fluorescence due to complex formation) with only one exception: K_b values were directly measured at all the temperatures investigated. The following values were obtained:

$$T^0[K] \quad 278.2 \quad 298.2 \quad 313.2$$
$$K_b[M^{-1}] \quad 6.55.10^5 \quad 2.36.10^5 \quad 1.05.10^5 \quad .$$

At the same temperatures, direct calorimetric measurements gave ΔH_b values listed below (in kcal per mol of NAD bounded).

$T^0[K]$	278	298	313
ΔH_b	-1.9	-12.4	-20.1
ΔS_b^0	+19.9	-16.8	-41.6

ΔS_b^0 values were calculated with the help of the same equilibrium formulas as in the previously discussed paper, i.e., assuming the validity of both postulates stated above and, consequently, the applicability of the Van't Hoff equation.

At the same time, Van't Hoff calculation of ΔH_b from K_b values gives $\Delta H_b = -9.0 \pm 1.0$ kcal per mole of NAD over the whole temperature interval studied. This value differs greatly from the true calorimetric ΔH_b values and from $\Delta H_b'$ value which, according to (5.33), is 143,08 kcal per 1 mole of NAD . This discrepancy makes us doubt the validity of the postulates on which the authors' calculations were based.

The fact that any local microchemical changes of a protein macromolecule (attachment of a low-molecular ligand, oxidation or reduction of a metal ion in the active center, etc.) can lead to conformational or configurational rearrangements of biopolymer molecules, can be considered now as having been established. Indirect evidence of this can be found in numerous works on the influence of these local changes upon physical properties (rotary dispersion, viscosity, solubility, sedimentation constants, etc.), on stability with respect to the action of proteolitic enzymes or denaturating agents. It is impossible to give all the references here (see, e.g. [5.25,41,42]). In recent years more direct proofs of the existence of such effects have been obtained due to extensive application of fluorescent and spin labels and probes.

It was, in particular, firmly established that the attachment of a substrate to the active center of the enzyme can result in essential changes in the environment of a label bound at a distant site of the protein globula (see, e.g. [5.43]).

DOBSON and WILLIAMS [5.44] have recently published beautiful results concerning conformational transitions in lisozyme induced by inhibitors (monomer, dimer and trimer of N-acetylglucoseamine). The authors applied the high-resolution NMR method (270 Mcps). They were able to resolve and identify NMR signals of aliphatic and aromatic protons of individual aminoacid residues. Analysis of the positions and widths of NMR signals showed the inhibitor attachment to be accompanied by substantial displacements of aminoacid residues localized at different sites of the polypeptide chain. Inhibitor attachment leads not only to the conformational changes of the lisozyme molecule but to the changes of its conformational mobility: the frequency of general and local spontaneous protein rearrangements decreases after inhibitor addition. Incidentally, the rates of unimolecular conformational transitions, following the bimolecular reaction of inhibitor attachment, are much slower than the rate of this attachment. These rearrangements take up to tens of milliseconds. Therefore, during one conformational transition a macromolecule must undergo many elementary acts of low-molecular inhibitor attachment and dissociation.

Results of several remarkable investigations have been published, where structural changes were directly recorded by means of X-ray diffraction. I am referring to PERUTZ's studies of structural rearrangements in a hemoglobin molecule after

oxygen additional to heme iron, and to the studies of DICKERSON et al. on the
structural changes accompanying the oxidation and reduction of iron in the cytochrome c heme.

5.4.1 Hemoglobin Oxygenation

Changes in hemoglobin molecule geometry, as well as in the electronic structure
of its iron porphyrin active centers (hemes), accompanying molecular oxygen attachment to the heme iron atoms (oxygenation) have been studied by many scientists
for many years. These changes have a direct relation to hemoglobin functioning, to
its ability to transfer large quantities of molecular oxygen at a given difference
between O_2 partial pressure in lung alveoles and tissues capillaries. Indeed, we
shall further see that just the specific conformational changes of protein globula
in the course of oxygenation lead to the cooperativity of the oxygen binding process and to the appearance of a specific S-shaped dissociation curve, providing a
considerable increase of transferred oxygen in comparison with the quantity that
would be transferred in the case of the classical Langmuir dissociation curve
(Fig.5.5). Therefore, the study of hemoglobin structure and dynamic characteristics
is connected with the solution of, probably, the most important contemporary
scientific problem: *the mechanism of purposeful behavior of biological systems at
the molecular level.*

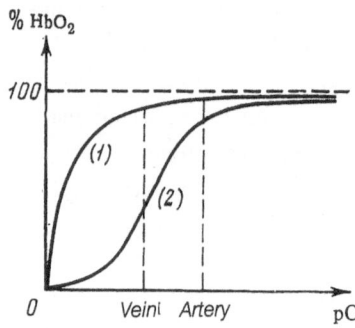

Fig.5.5. Hemoglobin hyperbolic curve and S-shaped
(curve 2) oxygenation curves

Discussions of historical, theoretical, and experimental aspects of hemoglobin
investigations can be found in various monographs and reviews (see, e.g. [5.45,46]).
In 1970 PERUTZ published two papers founded on his own X-ray studies and on some
literary data [5.47,48]. In these papers PERUTZ formulated the basic stages of geometrical changes in the hemoglobin molecule, which accompany the acts of oxygen
attachement in four subunits of this protein.

We shall now set forth the main conclusions which follow from PERUTZ's papers.
A molecule of deoxyhemoglobin differs from that of oxyhemoglobin by the tertiary
structures of its subunits, and by the quaternary structure as well, i.e., by the

mode of packin- of two α- and two β-subunits into the globula. Oxygenation begins with O_2 molecule attachment to the iron atom of one of α-subunits (the exact sequence of the oxygenation of heme units is not known and is not important for the present discussion). In α-subunits, the crevice between helical regions, where the heme ring is located, is broad enough, and the oxygen molecule can be attached to the heme iron, practically, without activation energy. The formation of the oxygen-iron bond results in the transition of the iron center from high-spin state with coordination number 5 to the hexacoordinated low-spin state. The iron atom, that was situated at a distance of 0.75 Å from the prophyrin ring plane, shifts and is now in this plane. The bond between the iron atom and the nitrogen atom of adjacent histidine residue becomes stronger, and the distance between this nitrogen atom and the porphyrin plane diminishes by 0.75-0.95 Å[6]. These shifts "trigger" subsequent conformational changes of tertiary and quaternary molecular structures. These changes are discussed in detail by PERUTZ. Conformational changes result in the displacement of one of the helical regions of the peptide chain in the direction of the globule center. As a result, tyrosin residue is pushed out from its cavity and two salt bridges between α_1- and α_2-subunits are ruptured. The same procedure occurs after α_2-subunit oxygenation, and two more bridges break down. This is accompanied by proton dissociation ("Bor-effect"). Thus, after the oxygenation of two α-subunits, four of the six salt bridges, determining the quaternary structure of deoxyhemoglobin, are ruptured, facilitating the severance of the two remaining bridges ($\alpha_1\beta_2$ and $\alpha_2\beta_1$). The tetramer assumes oxyconfiguration, and salt bridges connecting two β-subunits with 2,3-diphosphoglycerate molecule between them break down. 2,3-diphosphoglycerate is now expelled. At this stage, the macromolecule conformation is a mixed one: the quaternary structure is already of oxy-form, and the tertiary structure of β-subunits has not been changed yet but becomes strained. This leads to the facilitation of the attachment of next oxygen molecules, i.e., to the cooperativity effect and to the S-shape dissociation curve (if we had begun with the oxygenation of β-subunits the result, of course, would have been the same).

After that, the oxygen molecules combine with the iron atoms of β-subunits. The same events occur here, causing the changes in the protein globula tertiary structure (the shift of iron atom into the porphyrin ring planes, the diminishing of the distance between these planes and histidine residues, the displacement of certain helical regions, and the expulsion of tyrosine residues from corresponding crevices). The only difference is the necessity of appropriate thermal fluctuation for the first stage of the process to be ensured. Therefore, the first stage cannot be activationless.

---

[6]The change of the distance between histidine and the porphyring ring after oxygenation was predicted as early as 1952 [5.45,49] on the basis of indirect data. In 1966 HOARD [5.50] predicted that oxygenation must lead to the shift of the iron atom relative to the porphyrin ring.

It is difficult to overestimate the importance of these papers of PERUTZ, which have been described here only too concisely: I insistently recommend the reader to study these papers in the original. The irreproachable logic, the beautiful structural schemes will give you a true aesthetic pleasure (see also [5.51-53]).

The kinetics of conformational changes described above is of great interest. In order to measure it, one can use the ability of oxy- or carboxyhemoglobin to undergo dissociation and liberate O_2 or CO if illuminated by the light in the visible spectral region. The first stages of $HbO_2 \longrightarrow Hb + O_2$ transition were studied in [5.54]. An intense laser flash leads to the practically instantaneous dissociation of all oxygen (in the case of oxyhemoglobin) or all carbon monoxide (in the case of carboxyhemoglobin). The main changes of optical density due to dissociation come to an end in less than 10^{-9} s (time resolution of the method). The spectral characteristics of reduced hemoglobin, formed due to the O_2 (or CO) dissociation within less than 10^{-9} s after the flash, are, however, slightly different from the "normal" equilibrium hemoglobin: the optical density changes are too great. Transformation into the "true" hemoglobin requires about 10^{-7} s, and, after that, the spectrum does not undergo any further changes. It does not mean, however, that 10^{-7} s after the "instaneous" ligand (O_2 or CO) dissociation all hemoglobin transformations come to the end. Investigating carboxyhemoglobin photodissociation at pH 9.0 GIBSON [5.55] found that the hemoglobin, formed within 2.10^{-5} s after the flash, can combine oxygen considerably faster (by an order of magnitude) than the "normal" hemoglobin. The kinetics of the transition of "fast" to "normal" hemoglobin satisfies the first-order equation with the rate constant of about 200 s^{-1}, i.e., the duration of this transition is several milliseconds. This transition is not accompanied by spectral changes in the visible part of the spectrum. One can assume that fast changes occurring after ligand dissociation include only the prostetic group and its immediate vicinity. Heme are reduced, but the conformation of the rest of the macromolecule remains "oxygenated", because the corresponding conformational transition requires much more time than it takes for the changes in the heme electron structure to take place. Unfortunately, these results were not reproduced at neutral pH values.

The difference between the equilibrium conformational states of a macromolecule corresponding to different states of its active center becomes apparent not only in the molecular structure but in the changes of kinetic characteristics of these states, in particular, in their ability to undergo spontaneous structural fluctuations. It was shown as early as 1966 [5.56,57] that just these "conformational oscillations" (periodical reversible structural changes), as a rule, determine the rate of isotopic exchange of hydrogen atoms in proteins. ENGLANDER and ROLFE [5.58] have recently shown that, as a result of oxygenation, hemoglobin structure becomes appreciably more labile, and new large macromolecular segments containing about 30 hydrogen atoms, begin to participate in spontaneous "respiratory" movements.

The above kinetic pecularities of hemoglobin reaction with CO, namely, the comparatively slow structural changes of protein globula after fast changes of the active center caused by ligand attachment or dissociation, are, apparently, common properties of protein reactions. As a result of the first elementary act, there arises a nonequilibrium state in which the local changes in the vicinity of the active center have already taken place, but the main part of the macromolecule has preserved its original conformation. Such nonequilibrium states have been recently found and identified experimentally for several proteins, the reaction of ferri-hemoglobin electronic reduction included [5.59,60]. The reduction was carried out by thermal electrons formed by γ-irradiation of frozen solutions of ferrihemoglobin in water-ethyleneglycol mixture at 77 K. In these conditions, only the electronic recuction of iron and small changes in the vicinity of the active center are pos-sible, but the initial conformation of protein globula is kinetically stabile ("frozen"). In this way a steady nonequilibrium state of hemoglobin arises: the re-duced active center and the "oxidized" conformation. The reduced heme is now in a low-spin state (it is never in this state in equilibrium conditions) and has a "cytochrome-like" spectrum. The temperature increase and solvent defreezing lead to the relaxation of the molecule to the state of normal equilibrium reduced hemo-globin. Transient states (certainly, kinetically unstable) not unlike those out of equilibrium were recorded in the course of ferrihemoglobin electronic reduction of roon temperature in water solutions with the help of the method of pulse radiolysis [5.61]. For recent results see [5.109,110].

These and other results, indicative of a rather slow conformational rearrange-ment of protein macromolecules in the course of protein reactions, and of the ap-pearance of noticeable concentrations of essentially nonequilibrium transient con-formational states, compel us to regard some conventional ideas with caution. For many years hemoglobin served as a favorite model of protein allosteric interaction which is realized in macromolecular conformational changes. All kinetic schemes of these transformations always use invariable conformationally equilibrated protein molecules ("conformers"), and explain the changes in system properties only by changes of the relative concentrations of these conformers. This assumption is at the basis of all models of hemoglobin cooperative properties, beginning with the formal kinetic scheme of ADAIR [5.62] and ending with the model of "generalized concert transitions" devised by OGATA and MCCONNELL [5.63]. The latter is identi-cal to the model of MONOD et al. [5.64] in every respect, except that it takes into account the nonequivalence of α and β subunits. The authors of [5.64] assumed the existence of two types of equilibrium conformers T and R, having, correspondingly low and high oxygen (or CO) affinity. Oxygen attachment induces the shift of T$\rightleftharpoons$R equilibrium to the right which, in its turn, stipulates the cooperative properties of hemoglobin. A similar "equilibrium" approach is at the basis of the consecutive addition model [5.65].

High-resolution NMR experiments on CO binding by hemoglobin [5.66] have shown that, in the course of the binding process, there appear states which cannot be described using the concept of the equilibrium of conformers. The authors of [5.67-71] have come to the same conclusion in our laboratory. They studied the kinetics of transitions between the equilibrium (in the dark) and the steady (in the presence of light) states of the hemoglobin—CO system in a broad range of saturation degrees, hemoglobin concentrations and light intensities. It was shown that the results obtained cannot be, in principle, explained in the realm of any scheme involving only equilibrium forms of proteins with kinetic parameters of their reactions with ligands measured at equilibrium conditions. In the course of transition between the equilibrium and the steady state, the unequilibrated (not having enough time to relax) protein molecules appear with different kinetic characteristics.

5.4.2 The Oxidation and Reduction of Cytochrome c

Keen interest was aroused by the results of X-ray structural analysis of ferro- and ferricytochrome published in 1971 by DICKERSON and co-workers [5.72,73]. In these papers horse heart ferricytochrome c was compared with tuna heart ferrocytochrome c, and striking structural differences were revealed. The authors have recently renounced their results, and published new papers [5.74,75] where the comparison of tuna heart ferri- and ferrocytochrome c structures leads to the conclusion of the absence of any noticeable structural differences between the molecules in two redox states. Nevertheless, the results of [5.72,73] are well worth considering. The ideas on possible mechanisms of electron transfer through the cytochrome c globule formulated there stimulated a fruitful discussion and preserve their significance up to now, irrespective of their correctness.

In the membrane of mitochondria, cytochrome c realizes the electron transfer between the cytochromereductase (complex III, according to Green) and the cytochromeoxidase (complex IV) enzyme complexes. This protein with molecular weight of ~ 12300 contains 10_4 aminoacid residues in a single polypeptide chain without disulphide bridges. Figure 5.6, represents one of the stereoprojections of the ferricytochrome c structural scheme (according to [5.72]). The following covalent bonds exist between the heme and the protein: thioether bonds with cysteine residues 14 and 17 and bonds of iron atom with the nitrogen atom of histidine 18 imidazole (the fifth coordination site of iron) and with the sulphur atom of methionine 80 (the sixth coordination site). The imidazole ring of histidine 18 is also fixed by the hydrogen bond with the COO-group of proline 30. The fixation of heme in its "cavity" is also realized by two carboxyl groups of heme propionic acid side chains: one of these groups is connected by two hydrogen bonds through its oxygen atoms with tyrosine 48 and tryptophan 59, and the other protrudes through the molecular surface into the polar media. As it can be seen in the scheme, the polypeptide chain can be subdivided into two halves. Residues 1-47 are situated (in

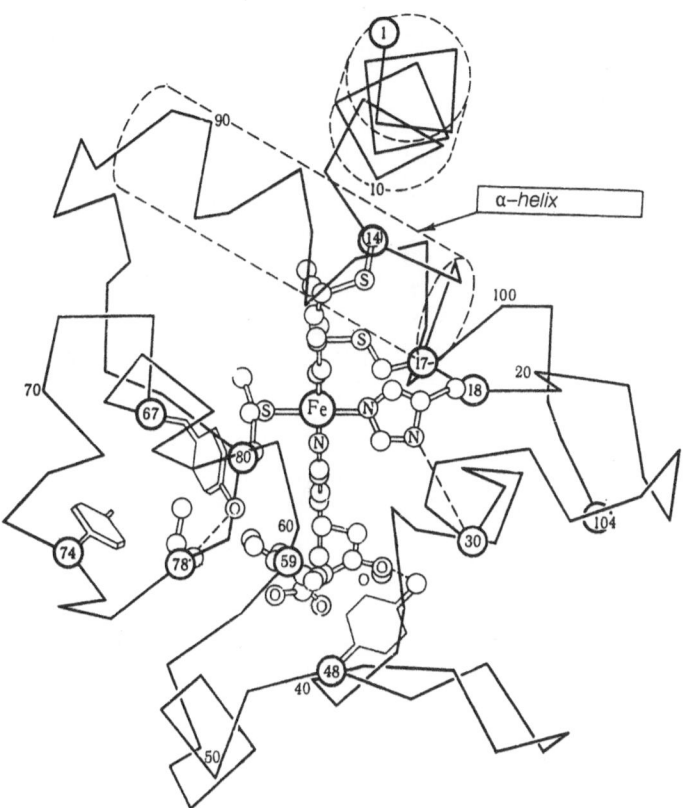

Fig.5.6. A stereoprojection of the ferricytochrome c structural scheme (according to [5.73])

Figs.5.6 and 5.7) to the right of the heme cavity, residues 48-91 to the left, and residues 92-104 form something like a briefcase clasp (through the top from left to right). The last two links of the polypeptide chain (103 and 104) appear to be fixed rather loosely and have some freedom of movement. The only part of the poly-peptide chain that forms an exact α-helix is the 91-101 section. The general struc-tural scheme of ferricytochrome c can be described as a polypeptide chain wrapped up around the heme with hydrophobic protein groups near the heme and polar groups at the surface. The globule has two "openings", which connect its inner parts with the surface. They are a channel to the heme cavity (to its left upper part) and a right channel, the bottom of which is fixed by the loop formed by residues 19-25. The right channel is available to solvent molecules and to small groups of the aromatic amino acid residue type. This channel, resumably, realizes the connection between cytochrome c and cytochrome oxidase.

The reduction of ferricytochrome c, i.e., the transfer of an electron to the iron atom, is accompanied by essential conformational changes (Fig.5.7). Only 21 amino acid residues, out of 104, i.e., the so-called molecular skeleton, formed by N-terminal helix (residues 1-11) and the region extending from 12 to 18 residues

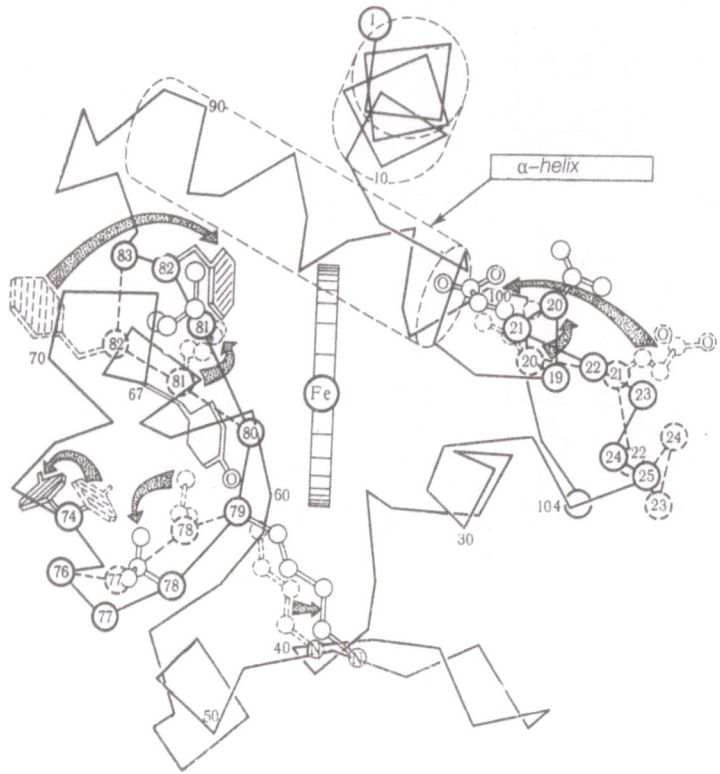

Fig.5.7. Main structural changes accompanying ferricytochrome c reduction [5.73]

which are a small section near the C-terminal, and the one near the heme itself
remain unchanged. The most significant displacements taking place during the reduction are listed below. The polypeptide chain section with residues from 77 to 83
undergoes the greatest shift. This section can be called the chain's "active
center", and it contains evolutionary invariant amino acids. Phenylalamine 82 in
the ferricytochrome molecule is shifted toward the far left edge of the globule
and leaves open the orifice leading to the upper left compartment of the heme
cavity. Phenylalamine in ferrocytochrome is moved to the heme cavity and closes
this orifice. The adjacent isoleycin 81 residue whose side group in ferricytochrome was situated at the globule surface, now turns and protrudes into the solvent. As a result, positions of residues 77-79 are also changed.

   In ferricytochrome, chain 55-75 forms a loop around the hydrophobic, inacces-
sible to the solvent, region (the left channel), where the aromatic rings of
tyrosine 74 tryptophan 59, and tyrosine 67 are situated. These rings are parallel
to each other and, presumably, form an electron path for the electron transfer
from cytochromereductase through the S-atom of methyomine 80 and to the heme iron
atom. Tyrosine 67 is fixed by the hydrogen bond with treonine 78. However, in the
process of cytochrome reduction this bond breaks down, the tyrosine 74 ring plane

turns by about $60°$, and is now almost perpendicular to the globule surface (in ferricytochrome this plane was practically parallel to the surface). Furthermore, all of the adjacent polypeptide chain changes its position[7].

In the right channel (residues 19-25) the reduction is also accompanied by displacements closing this channel.

The conformation of ferrocytochrome is, thus, much more dense and "closed" than the conformation of this protein in ferriform. As has already been stated above, this interpretation of X-ray data was given on the basis of the investigation of horse heart ferricytochrome C and tuna heart ferrocytochrome C in crystalline states [5.72,73]. Having compared the structures of tuna heart ferri- and ferrocytochrome C, the same authors concluded that there are no significant conformational or configurational changes accompanying the reduction of ferricytochrome C [5.74,75]. This conclusion is rather surprising. Numerous investigations of ferricytochrome C and its reduction in solution are indicative of considerable structural changes (see, e.g. [5.76]).

ULMER and KÄGE [5.77] studied the hydrogen isotopic exchange rate in the cytochrome c molecule for two oxidation states of a heme iron. Iron oxidation was shown to result in an essential increase in the number of fast exchangeable hydrogen atoms. It follows from the above, that the ferricytochrome structure is not only more "open", but also that ferricytochrome is more labile, that the protein globule has greater amplitudes of fluctuation oscillations.

It should be borne in mind that practically all studies in solution were carried out with horse heart cytochrome C. In our laboratory it was shown that the amplitudes of structural changes in the active center region, accompanying the electron reduction of ferricytochrome C, can vary greatly, depending on the source of cytochrome [5.78]. It is possible that cytochrome C of tuna is characterized by specially small structural changes during redox transformations. On the other hand, it is possible that these changes in crystalline state differ essentially from those in solution.

Estimations of the rate of cytochrome C conformational changes after its reduction were made in important work [5.79], dealing with the kinetics of ferricytochrome C spectral changes, accompanying its transformation into ferroform under the action of hydrated electrons obtained by pulse radiolysis. Iron reduction is completed sooner than after 10^{-6} s, but the spectrum obtained differs from that of equilibrium ferrocytochrome C. Further changes of the spectrum, up to its total coincidence with the usual ferrocytochrome C spectrum, require a slow monomolecular process with $k \approx 8.5$ s^{-1}. Whether these changes have time to take place in the course of cytochrome C normal functioning is not known. Conformational changes, accompanying cytochrome C hydration, require even more time [5.80].

---

[7]Of course, the electron path breaks down after that. It is, however, possible that this path is not of any importance for electron transfer that is realized in accordance with the tunnelling mechanism (see Chap.7).

Nonequilibrium conformational states of cytochrome C arising in the course of its reaction have been recently studied in our laboratory. During the low-temperature reduction of ferricytochrome C in the rigid matrix by thermal electrons there appear stable (at low temperature in frozen matrix) intermediate states (the active center is reduced, the main part of the protein globule has an "oxidized" conformation) [5.78]. Cytochrome C reduction with conventional chemical agents at room temperatures in water solutions seems to involve similar intermediate conformationally unequilibrated states. This was shown in the study of complex formation between cytochrome C and CO [5.81]. Equilibrium ferricytochrome C does not bind CO because a complex can only be formed with the reduced heme. Equilibrium ferrocytochrome C, practically, does not bind CO either due, probably, to the fact that a "closed" protein conformation does not allow the CO molecule to approach the iron atom. At the same time, a complex is easily formed when the reduction is carried out in the presence of CO. Cytochrome binds CO in the intermediate conformational state when iron is already reduced but the globule conformation did not have enough time to change.

It was also shown that, in the conditions of dynamic redox equilibrium in solution, there are no conformational transitions between the equilibrium oxidized and reduced forms of cytochrome C (these transitions do not have enough time to take place).

The intermediate conformationally out of equilibrium states of ferricytochrome C were also recorded after fast changes in pH or the ionic strength of solution [5.82,83].

New data concerning cytochrome C out-of-equilibrium states and their physical and chemical properties can be found in [5.109-113].

5.5 On the Compensation Effect

For many years, when discussing the thermodynamics and kinetics of processes taking place in the condensed phase (chiefly, in solutions), people pondered over the so-called compensation effect. This effect is particularly evident in biopolymer processes (denaturation, complex formation, inactivation, enzyme catalysis). In this paragraph, where the main experimental facts concerning the compensation effect will be summed up and in the next paragraph, where I am going to try to analyze and clarify the physical meaning of the parameters which are determined in the kinetic and thermodynamic studies of such processes, we shall be dealing with biopolymers. I think, however, that all the conclusions reached are also valid for the case when the compensation effect is observed with low-molecular compounds.

Phenomenologically, what one understands by the thermodynamic compensation effect is the existence of a linear dependence between enthalpy (ΔH_i) and entropy

(ΔS_i) changes in the set of related processes

$$\Delta H_i = \alpha + T_c \Delta S_i \quad (i = 1,2, \ldots) \quad . \tag{5.36}$$

where i determines the process, and α and T_c are constants. The constant α is, evidently, the free energy of any process in the given set at $T = T_c$. The constant T_c has the dimension of temperature and is usually called the *compensation tempera-ture*. All processes of the set have, thus, equal equilibrium constants $K = \exp(-\alpha/RT_c)$ at the same temperature $T = T_c$. The plots of compensation effect are usually drawn either in coordinates ΔH_i - ΔS_i (Fig.5.8a), or in coordinates $\Delta\Delta H_i$ - $\Delta\Delta S_i$ (Fig.5.8b), where $\Delta\Delta H_i$ and $\Delta\Delta S_i$ are the differences between $\Delta H_i (\Delta S_i)$ and the corresponding values for a member of the set that was chosen as a standard (usually the role of the standard is played by the reaction with the lowest ΔH_i and ΔS_i values). The second way of presentation corresponds to equations

$$\Delta\Delta H_i = T_c \Delta\Delta S_i \quad . \tag{5.37}$$

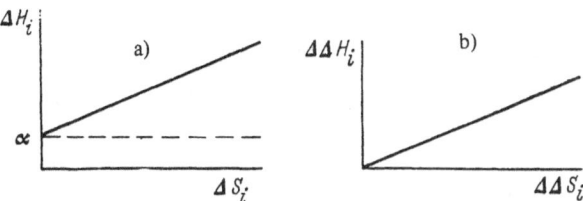

Fig.5.8 a,b. Two ways of the presentation of compensation effects

In the same way we can formulate the kinetic compensation effect. In this case, activation enthalpy (or energy, see Sect.5.3), ΔE_{ai}, activation entropy, ΔS_{ai}, and activation free energy ΔF_{ai} are substituted for the enthalpy, entropy and free ener-gy of the process. For the kinetic compensation effect we have

$$\Delta E_{ai} = \alpha + T_c \Delta S_{ai} \quad (i = 1,2, \ldots) \quad . \tag{5.38}$$

Instead of equilibrium constants there are now equal rate constants for all the processes in the set at $T = T_c$. Reviews of experimental data and theoretical con-siderations related to the thermodynamic and kinetic compensation effects can be found in [5.84,85].

There are several ways to construct a set of reactions for compensation effects, i.e., methods of choosing the values of ΔH_i, ΔS_i to draw compensation plots shown in Figs.5.8a,b. The first method consists in using the regular chemical structural changes of one of the reagents to obtain a homologous set of similar chemical pro-cesses. The second method consists in using the regular changes in reaction medium (changes of pH, electrolyte concentrations, ethanol content, etc.).

All compensation effects described in the literature can be subdivided into two types. The first type includes those effects for which the T_c constant is very close to the temperature of experiment (the so-called total compensation effect). Typical examples of such effect can be found in [5.84,86] on the thermodynamics and kinetics of protein denaturation and the kinetics of enzyme reactions. In these cases, T_c values were quite close to the temperatures of measurements, which, in their turn, may vary within a narrow interval only, due to the specificity of the objects. Points on compensation plots have a large scattering, and the really measured (without the use of Van't Hoff or Arrhenius equations) equilibrium or rate constants differ but slightly from one another (in the case of a greater difference, they could not be measured with the help of standard methods). At these conditions, the compensation effect is inevitable[8]. The situation is really quite simple. There is a set of close values of, for example, equilibrium constants, K_i, which are represented in the form of products of two factors

$$K_i = \exp(\Delta S_i/R)\exp(-\Delta H_i/RT) \quad . \tag{5.39}$$

It is easy to understand that, if the products of these factors are to be close to each other and measurement temperatures cannot differ greatly, relation (5.39) must hold true with the factors chosen arbitrarily.

In this case, the question is not what is the nature of compensation, but why can similar processes with close-lying equilibrium or rate constant values (and only these values are actually measured) differ so greatly in ΔH and ΔS (or ΔH_a and ΔS_a) values, calculated with the help of Van't Hoff or Arrhenius equations. It should be borne in mind that these differences can reach the values of hundred of kilocalories or entropy units.

The second type of effects can be illustrated by high-precision measurements giving compesnation plots with a small scattering of points and with T_c values that do not coincide with the measurement temperature. There are many examples of this type in the review [5.85] quoted above. The works of BEETLESTONE et al. concerning the thermodynamics of some vertebrate hemoglobins binding with various ligands [5.87-91] may serve as an example.

To obtain data for his compensation plots, the researcher must necessarily use equations of the Van't Hoff or Arrhenius type, and, consequently, assume the validity of all postulates on which these equations are based. In the case of kinetic compensation effect there are no other methods to find ΔS_{ai} and ΔE_{ai} values. In the case of the thermodynamic compensation effect, there have recently been developed direct calorimetric methods of ΔH_i measurements. However, to determine ΔS_i one must, nevertheless, carry out the treatment based on the validity

---

[8]This was stated by Professor I.V. Berezin.

of the Van't Hoff equation and, consequently, that of equilibrium thermodynamic relation (5.3).

We shall not discuss here the existing theories of compensation effects. The reader, to whom it is of interest, can find the needed literature in the above-mentioned reviews. It is probably already clear now that, from my point of view, the compensation effect is not a physical phenomenon but the result of improper application of some equations of equilibrium thermodynamics.

5.6 On the Validity of the Van't Hoff and Arrhenius Equations and of the Activated State Theory for Biopolymer Reactions

It is convenient to discuss the thermodynamic and the kinetic parameters separately. We shall begin with thermodynamics. Let us sum up the basic conclusions which follow from the analysis of the experimental data presented in this chapter.

1) The structure of globular protein in solution, in the temperature region below denaturation (0-40°C), changes with temperature. These changes may be accompanied by heat absorption.

2) The complex formation between low-molecular compounds and protein, and other local microchemical perturbations leads, as a rule, to considerable rearrangements of the protein globule.

3) The protein reaction (denaturation, complex formation) heat effects measured calorimetrically have a strong temperature dependence (up to the change of heat effect sign) at comparatively small temperature changes.

4) The heat effects of the protein reaction, determined by means of the Van't Hoff equation, have, as a rule, a weak temperature dependence, except for the abrupt changes in calculated ΔH values (the so-called "breaks" on $\lg K$ - $1/T$ anamorphoses) observed at certain temperatures. The "equilibrium" ΔH values differ, sometimes greatly, from the true calorimetric values.

5) If protein transformation is caused by temperature change (e.g., thermal denaturation), treatment by means of the Van't Hoff equation gives ΔH values lying close to the true values.

Let us consider now the nature of temperature dependence of true, calorimetrical measured heats of protein reactions. This dependence is usually explained in the literature by the heat capacity difference between the initial and the final state. This explanation does not stand up under examination. Indeed, if it was correct, then, for instance, the intrinsic ΔH° value (at 0 K) of the binding reaction between NAD and glyceraldehyde-3-phosphate dehydrogenase would, according to the above data, be equal to 143 kcal/mol^{-1} (Sect.5.4). The same value for the binding of hexytol-1,

6-diphosphate with aldolase would be 329 kcal/mol^{-1} (Sect.5.4), and for acid denaturation of ferrihemoglobin 2900 kcal/mol^{-1} [5.10]. The absurdity of these
values is self-evident. The explanation of temperature dependence in terms of the
Kirschhoff law would imply the applicability of thermodynamic relation (5.3). This,
in turn, as has been shown above, would lead to equality between reaction enthalpy
values measured calorimetrically and by "equilibrium" methods. This equality does
not hold true. Thus, apparently the second explanation of the temperature dependence of biopolymer reaction enthalpy values given above (see Sect.5.2) is valid.

How does a protein reaction proceed? Let us initiate a protein reaction at
temperature T_1 (the initiating agent may be, for instance, a low-molecular compound
forming a complex with protein, or an acid for acid denaturation, etc.). The reaction is, as a rule, accompanied by essential structural transformation of the protein. SHIAO and STURTEVANT's point of view, according to which the calorimetrically
measured enthalpy values for inhibitor binding to α-chymotrpysin reflect, mainly,
the protein conformational changes, is probably true for most protein reactions.

Now let us initiate this reaction with the same agent at a different temperature,
T_2. The transition to the new temperature would lead to a change in the initial and
the final protein structures[9] and not only to the change of its equilibrium structures, or, using the chemical language, to protein isomerization. A constructional
change can result in a change of the possible paths of protein transformation induced by the initiating agent. It may lead to the disappearance of some and to the
appearance of other mechanical degrees of freedom. After the perturbation, parts of
protein construction will move differently and relax to the new quasi-equilibrium
state. Thus, at temperatures T_1 and T_2 different compounds would react in a differ-
rent way. As stated above, in this case the temperature dependences of the overall
reaction enthalpy and entropy would not obey (5.3).

What does the scientist measure when he determines, as in this case, the "equilibrium constant"[10] at different temperatures? He determines the temperature dependence of initial states and reaction paths, rather than the temperature dependence
of the "equilibrium constant". The ΔH and ΔS values obtained using the Van't Hoff
equation are not enthalpies and entropies of any definite transformations, because
they are calculated assuming the validity of postulate (5.3) that does not hold true
in reality. For the processes considered, these quantities have no strict physical
meaning. To be more exact, such a meaning (specific for each process) could be

---

[9] In principle, these structures may remain unchanged but experiments show that,
in most cases, within the temperature range of 5-50°C, where biochemical investigations are usually carried out, the structural rearrangements do take place.

[10]The words "equilibrium constant" are in quotation marks not only because we are
dealing with quasi-equilibrium (in Lifshitz' sense, see Chap.4) systems with
rigid memory, but also because this memory, i.e., the system construction, changes
with temperature.

introduced, if all conformational, configurational, and enthalpy changes of the
protein with temperature and those taking place after agent addition were known.
But, in this case, no knowledge of "equilibrium" values of ΔH and ΔS would be
needed.

The fact that temperature dependence of the measured equilibrium constant is
exponential (i.e., the Van't Hoff equation is formally valid, even with constant
values of ΔH and ΔS) means nothing. There are not so many complex processes proceeding within a narrow range of parameters that could not be described with exponential functions.

A different situation arises when the initiating agent is the temperature as
such, as for protein thermal denaturation (see, e.g. [5.92]). Let us consider this
procedure at greater length. In this case, the transformation temperature for a
given protein is determined by ionic strength (μ) and pH. Measurements at different temperatures are, essentially, those performed at different μ and pH values.
The calorimetrically determined heat of denaturation is (at fixed μ and pH values)
the enthalpy difference at temperatures lower and higher than that of the start
and the completion of transition, respectively. Transformation takes place within
a comparatively broad temperature range (5-10°C), and transition temperature is
usually understood as the midpoint of this range (if the transition is symmetric).
If the transition is not symmetric, it will be "adjusted" by relating some parts
of transition to the pre-denaturation changes or by extrapolating the basic line
"with normal heat capacity changes" to the transition region (see Sect.5.2). The
equilibrium constants will be obtained here by determining the "transition sharpness" in the temperature range where denaturation is the only structural transition
realized. If this transition is symmetric, it proceeds as if the transformation
had taken place at a fixed temperature corresponding to a 50% transition. Therefore,
as was rightly stated by Privalov, the coincidence of "equilibrium" and calorimetric
results means, in this case, only that one structural transformation of the phase
transition type occurs during heat denaturation (the treatment of results is such
that all other structural changes can be neglected!). Under these conditions, transition entropy can be calculated using the classical expression

$$\Delta S = \Delta H/T \tag{5.40}$$

where T is the transition temperature in the meaning formulated above. The temperature dependence of enthalpy and entropy of the heat denaturation process can, however, hardly be discussed using the experimental results obtained with different
solvent compositions (for, instance, pH) and, consequently, with different transition temperatures. The pH (or μ) changes, as the temperature changes in previous
examples would lead to macrostructural changes of the protein in its initial and
final states. Measurements at two different temperatures, T_1 and T_2, corrrespond
again to different systems and different processes. For each temperature relation
(5.40) holds true. Changes in ΔH and ΔS values, as one passes from one transfor-

mation temperature to another, do not, however, reflect the temperature dependence of these parameters but are associated with the complex changes of the system construction and reaction path due to pH, μ, and temperature changes. If perturbations are small and transition temperatures differ but slightly, i.e., if

$$|T_1 - T_2| \ll T_1 \qquad (5.41)$$

then (5.40) automatically leads to[11]

$$\Delta\Delta H_i \approx T\Delta\Delta S_i \qquad (5.42)$$

which is formally equivalent to (5.37) for the compensation effect with compensation temperature T_c close to transition temperatures T.

Thus, the correlation between $\Delta\Delta H_i$ and $\Delta\Delta S_i$ values, sometimes noted in this kind of investigations is of trivial character and has nothing in common with the compensation effect discussed above.

Let us consider now the question of the applicability of Arrhenius and activated state theory equations. As has already been mentioned, application of these equations is equivalent to the assumption of a thermodynamic equilibrium between the initial state of the system and the state of the activated complex at the top of the activation barrier and to the assumption that thermodynamic relation (5.3) is fulfilled. For kinetic activation parameters this equation can be rewritten as

$$\frac{\partial\Delta E_a}{\partial T} = T \frac{\partial\Delta S_a}{\partial T} . \qquad (5.43)$$

It is clear that all the arguments, used to show that these postulates do not hold true for the thermodynamic treatment of protein reactions, may be applied to the kinetics of these processes as well. In the case of thermodynamic parameters, however, there is always a possibility of direct measurements of reaction heats. In the case of activation parameters, there is no such possibility. Therefore, one must be very careful with the values of activation parameters. The difference in heat content between the initial state and the state at the top of the activation barrier may change with temperature not only due to the heat capacity difference in these states, but also due to the changes in the barrier as such, and to the changes in reaction paths.

As we have already seen in this chapter (and shall see in the next one, when considering the details of the mechanisms of enzymatic processes), the behavior of globular protein during a reaction resembles the operation of a mechanical device rather than the course of a homogeneous chemical reaction. Therefore, small constructional changes taking place with temperature may result in changing the ac-

---

[11] $\Delta H_1(T_2 - T_1)/T$ being neglected.

tivation barrier and the reaction path without corresponding   as would be expected from (5.43)   changes in activation entropy.

Indeed, what is the behavior of the "activation barrier" in a clock mechanism with different thermal expansion coefficients of its components. It is clear, that its changes with temperature are completely determined by its construction. For instance, at some low temperature two gears do not fit, and the rate will be low. With a temperature rise, the adjustment will improve and the rate will increase. At a higher temperature the fit will again be poor (the gears become wedged), and the rate will decrease. If we try to describe this last step by means of a relation of Arrhenius type, the "activation energy" obtained would be negative.

It is clear that the same kind of reasoning may be applied to the formation of a complex between globular protein and a low-molecular ligand. Let us suppose that two ligand groups are bonded to two different aminoacid residues that become neighbors in space due to the polypeptide chain folding into the secondary and the tertiary structures. It is quite a common situation. Let us also suppose that the kinetics of the process is recorded by means of a method which does not feel the subsequent (after binding) conformational relaxation of a protein molecule. Even in this "classical" case the true activation barrier of the binding process (i.e., of the first stage of transformation) may be essentially dependent on temperature. Indeed, the temperature within the limits of its changes in biochemical research does not, virtually, influence the low-molecular ligand geometry, but its variations may lead to changes in protein globula configuration and, consequently, in the distance between the binding sites on the protein. Due to the strong dependence of chemical interaction energy on distance, the above would induce a notable change in the activation barrier even with very small configurational changes which do not affect the system entropy.

Naturally, the same situation may be realized in the case of one-site ligand binding (due, for instance, to steric hindrances).

What will the treatment of experimental data by means of classical equations (5.22,23) give in these cases?

Let us assume that the true activation barrier, ΔE_a, changes slightly with temperature without any relevant [according to (5.43)] changes in the true activation entropy, ΔS_a [12]. In the narrow temperature range (usual for the study of protein reactions) any weak dependence may be approximated by a linear one

$$\Delta E_a = \Delta E_a^0 + bT \tag{5.44}$$

where ΔE_a^0 and b are constants. Introducing (5.44) into (5.22,23), we obtain

---

[12]Such an assumption was made as early as 1961 [5.93] to explain the compensation effect observed in the case of the kinetics of free radical recombination in solids.

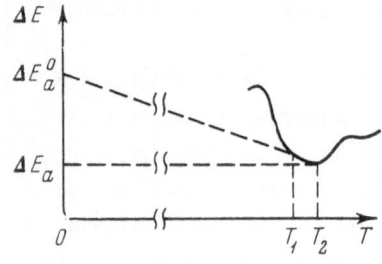

Fig.5.9. True, ΔE_a, and measured, ΔE_a^0, activation energies

$$\kappa = A \exp(-b/R)\exp(-\Delta E_a^0/RT)$$

$$= \frac{kT}{h} \exp\left(\frac{\Delta S_a - b}{R}\right)\exp(-\Delta E_a^0/RT) \quad . \tag{5.45}$$

Thus, in this case, the experimental data will satisfy the Arrhenius equation, but we shall be measuring not the true activation parameters ΔE_a and ΔS_a but the effective quantities of $\Delta E_a^0 = \Delta E_a - bT$ and $\Delta S_a^0 = \Delta S_a - b$ that may differ greatly from the true values. From (5.44) it is evident that the measured value, ΔE_a^0, is obtained by extrapolating the weak temperature dependence of ΔE_a to 0 K (Fig.5.9). If, for instance, the true activation energy changes by 1 kcal/mol^{-1} when the temperature rises from 20 to 30°C, the measured value would differ from the true one by 30 kcal/mol^{-1}. This corresponds to the exponential factor changing by 22 orders of magnitude. Using in this case the Arrhenius equation we cannot determine the true activation energy and cannot even be sure, whether the calculation was erroneous or not!

Let us introduce the difference $\Delta S_a - \Delta S_a^0$ instead of b into (5.44). After some arithmetic we obtain

$$\Delta E_a^0 = \Delta E_a - T\Delta S_a + T\Delta S_a^0 = \Delta F_a + T\Delta S_a^0 \quad . \tag{5.46}$$

Thus, if in the chosen set the true activation parameters (i.e., the true reaction rate constants) are changing but slightly, and measurements are carried out in a narrow temperature range, the activation energies and entropies obtained by means of the Arrhenius equation are interconnected by (5.38), constants α and T_c being the reaction free energy and the measurement temperature, respectively.

Such considerations can explain the compensation effect with $T_c \approx T$ and the enormously wide range of the measured parameter values in the compensation plots. Let us now see in what way can be explained the compensation effect of the second type, with $T_c \neq T$, if we do not want to restrict ourselves to general statements that our compensation plots were drawn using the data obtained from equations based on postulates which are not fulfilled.

Let, at some fixed conditions (pH, solvent, etc.), the true (but unknown to us) activation parameters of some protein reaction be ΔE_{a1} and ΔS_{a1}. Let us introduce

a small change into these conditions. Now, strictly speaking, in situations 2,3,...,
i,...,n other compounds will undergo reactions proceeding along other paths. However, if the changes are small (and the compensation sets are actually chosen in
this way), the true parameters will change but slightly, and we may approximately
assume that these changes are linearly interconnected

$$\Delta\Delta E_{ai} = T_c \Delta\Delta S_{ai} \tag{5.47}$$

where T_c is a constant having temperature dimensionality the exact value of which
depends on the construction of the system and on the type of changes introduced. At
a given temperature T

$$\Delta E_{ai} = \Delta E_{a1} + \Delta\Delta E_{ai} \quad , \quad \Delta S_{ai} = \Delta S_{a1} + \Delta\Delta S_{ai} \tag{5.48}$$

and the true free energy of activation, F_{ai}, is

$$\Delta F_{ai} = \Delta E_{ai} - T\Delta S_{ai} = \Delta E_{a1} + \Delta\Delta E_{ai} - T\Delta S_{a1} - T\Delta\Delta S_{ai} \quad . \tag{5.49}$$

If $T = T_c$, then [see (5.47)]

$$\Delta F_{ai} = \Delta E_{ai} - T_c\Delta S_{ai} = \Delta E_{a1} - T_c\Delta S_{a1} = \mathrm{const} \quad (i = 1,2, \ldots, n) \quad . \tag{5.50}$$

Thus, if equality (5.47), which is approximately true in the case of small changes
in the reaction set investigated, is satisfied, there must exist a certain temperature $T = T_c$ at which all reactions in the set have the same rate constant

$$\kappa = \frac{kT}{h} \exp(-\Delta F_{ai}/RT_c) \quad . \tag{5.51}$$

All the above relates to unknown true activation parameters undergoing but small
changes in the set 1,2, ..., i,...,n. The compensation "paradox", however, consists
not in the fact that there is a linear dependence between small changes of ΔE_a and
ΔS_a in the set of homological transformations, but in the existence of an enormous
range of "observed" changes in the measured activation parameters in such a set. We
must bear in mind that, by investigating the temperature dependence of each reaction rate and using (5.22,23), we obtain not the true parameters ΔE_{ai} and ΔS_{ai}, but
quantities ΔE_{ai}^0 and ΔS_{ai}^0 which have no clear physical meaning. Using the same
reasoning as in previous pages we obtain that the following compensation equation
holds for these "observed" fictitious quantities:

$$\Delta E_{ai}^0 = \Delta F_a^0 + T_c\Delta S_{ai}^0 \quad , \tag{5.52}$$

where constant ΔF_a^0 is the true free activation energy of any of the n reactions at
compensation temperature.

The compensation effect may, thus, be observed, when perturbations accompanying the
transition from one reaction of the set to another, are small, and when (5.43) is
not fulfilled and, consequently, the use of the Arrhenius equation and of the

transition state theory is meaningless. The same considerations may be, in principle, applied to the thermodynamic compensation effect.

Of course, the above cannot be regarded as a theory of the compensation effect. These arguments help to understand the reasons for the appearance of the relations of the (5.36) or (5.38) type, interconnecting parameters, which do not possess the physical meaning usually attributed to them. As has already been stated, in the realm of ideas presented here the compensation effect does not exist as a physical phenomenon. We can even maintain that finding such effect in the cases when Van't Hoff or Arrhenius equations are used for the treatment of experimental data suggests that these equations cannot be applied, and that parameters obtained have no clear physical meaning. Moreover, the "breaks" often observed on the Arrhenius and Van't Hoff anamorphoses cannot be regarded as evidence of structural rearrangements occurring just at the temperatures of these breaks. Configurational changes of a macromolecule may proceed gradually over some temperature interval, but at the "break temperature" the b values (5.44) change abruptly (for instance, when the changing distance between the binding sites on the macromolecule passes through the value corresponding to the distance between the binding sites on the low-molecular ligand). In order to interpret such a break, one must know in detail the construction of the system, its changes with temperature and those taking place in the course of the process in question.

All the above seems also to be true for many processes in the condensed phase involving no biopolymers. Such structures as solvate shells would, in this case, act as "macromolecules".

5.7 On Spontaneous Conformational Oscillations of Protein Macromolecules

It was mentioned in Chap.4 that, at certain conditions in protein macromolecules, one must observe spontaneous transitions between the close-lying conformational or configurational states. There have been publications concerning several types of investigations, where such "conformational oscillations" were, probably, observed. First of all, we must, mention extremely interesting works by SHNOLL and his co-workers carried out for more than 15 years [5.94-100]. These works present evidence that in the freshly prepared actomyosin and myosin solutions such characteristics as enzymatic activity, the number of available surface SH-groups, the ability to adsorb certain dyes undergo oscillatory changes with a frequency not higher than 10 c/s. Similar effects were found by CHETVERIKOVA for creatine kinase [5.101-103]. The summary of these experiments and some new data have been recently published in [5.104]. The authors consider their results to be indicative of the existence of spontaneous conformational changes in protein macromolecules. The most striking feature of these oscillations is their synchronism in macrovolume (otherwise they could not be detected with the methods applied). There is up to now no more or less verisimilar explanation of the origin of these synchronous oscillations.

There have been investigations of another type, where certain peculiarities of proton magnetic resonance spectra of certain proteins and polypeptides are interpreted as indicative of the existence of conformational oscillations with frequencies from tens of c/s (ribonuclease) to tens of kc/s (certain polypeptides [5.105-107]. This interpretation is based on the effective averaging of local magnetic fields on certain hydrogen nuclei during conformational oscillations. We must also not forget the old works of LINDENSTRÖM-LANG [5.108], who showed the rapid isotopic exchange of peptide hydrogen atoms in the inner portion of the globule to be indirectly indicative of conformational oscillations.

I cannot discuss here in detail the data concerning conformational and configurational oscillations of protein globules. If these data and their interpretation are confirmed, they open up a new extremely important stage in the development of biological physics.

5.8 Conclusions

The contents of this chapter allow one to formulate the tasks confronting the scientists who are investigating conformational and configurational changes in biopolymers. These tasks are:

a) Direct recording of the structural changes of macromolecules in the course of their reactions.

b) Direct recording of the structural changes of macromolecules with temperature.

c) Experimental fixation of rapid and slow (relaxational) stages of the above-mentioned transformations.

d) Development of a theory that will give experimentalists the methods of interpreting their results instead of using thermodynamic and kinetic methods that have been worked out for homogeneous gas reactions and are inapplicable in this case.

All these tasks are much more difficult to accomplish than it is to measure the temperature dependence of equilibrium and rate constants, to plot logarithmic anamorphoses and to interpret results in terms of system ordering or disordering according to the sign of calculated ΔS values.

6. The Physics of Enzyme Catalysis

6.1 Background

Perhaps the most important function of proteins is their catalytic, enzyme function. It is quite hopeless to try to present here a more or less comprehensive review of established experimental facts, empirical rules, and proposed theories concerning enzyme catalysis. An enormous (probably, even redundant) number of monographs and reviews have been published dealing with this question. Some of them are listed in [6.1-5].

All enzymes are proteins, and the whole protein molecule is necessary for the regular functioning of an enzyme[1]: we can, therefore, suggest that all that was written in Chap.5 about the pecularities of processes involving biopolymers will be also true for the chemical reactions catalyzed by enzymes. Indeed, statements on the importance of conformational lability of proteins in, for instance, the regulatory mechanisms of enzymatic catalysis have recently been transferred from the category of "it is impossible" to the category of "it is trivial", having passed rather hastily the stage of experimental and theoretical analysis.

In this chapter we shall try to analyze the consequences of applying to enzymatic processes the conclusions reached in Chap.5 in the course of discussing the thermodynamics and kinetics of protein conformational and configurational changes. The chapter is designed according to the following plan. First, general principles of the existing theoretical approaches to explaining the high activity and other properties of biocatalysts will be outlined. After that, we shall discuss experimental data that evidence the necessity of extending the conclusions of Chap.5 to the enzyme reactions. The "machine-like" nature of enzyme functioning will be illustrated by several more or less explicitly studied enzyme processes. The final part of the chapter is devoted to the formulation of certain new ideas on the physics of the elementary act of enzyme functioning.

---

[1]Experimental results, according to which the activity of certain enzymes is preserved after fragmentation [6.6,7], proved to be erroneous [6.8,9].

6.2 Existing Interpretations of Enzymatic Activity

First of all, a few words about one delusion that is quite common among biochemists.
Many people think that there exist certain general theories of catalysis in chemistry, but the application of these theories to enzyme reactions is difficult due
to the complicated nature of biochemical processes and biological catalysts. This
erroneous point of view is founded on the biochemists' unjustified worship of the
power of classical physical chemistry. Unfortunately, there is no such thing as a
more or less consistent general theory of catalysis.

One can find a lot of experimental data giving evidence that, in many cases,
enzymatic reactions obey the same well-known empirical and semi-empirical rules as
those governing the homogeneous catalytic reactions in solutions (acid-base catalysis, catalysis by metal ions and complexes, etc., see, e.g. [6.1]). In homogeneous
catalysis we are dealing, essentially, with the problem of reaction kinetics in the
liquid phase. The catalyst here is only one of the reagents that forms an intermediate compound and allows the overall reaction to proceed along a new path. The
physical theory of liquid phase reactions has not been as yet created. Thus, the
usual homogeneous chemical catalysis does not possess any advantage (from the
theoretical point of view) over enzymatic catalysis.

If we look at the theoretical aspects of heterogeneous catalysis, the state of
affairs here is even worse than in the case of enzymatic catalysis. It is, first
and foremost, due to the fact that, in the case of enzymatic catalysis, it is much
easier to perform a pure and reproducible experiment. However complicated the
enzyme would be, however careful one must be during its preparative isolation, an
enzyme is, after all, the sum total of macromolecules that have been synthesized
within the cell on a matrix and are practically identical. Therefore, the reproducibility of the functioning of isolated enzymes is, as a rule, much better than
the reproducibility of the activity of solid catalysts of abiogenic origin, whose
fine structural features, which determine their catalytic power (their surface
defects, etc.), are almost impossible to reproduce.

Before speaking about the interpretations of enzymatic activity it is necessary
to define more clearly what it is that requires interpretation. In other words,
what is the meaning of the statement: "a chemical reaction is hastened by an enzyme". It is not enough to say that the rate of the studied enzymatic reaction
must be compared with the rate of the same reaction without the enzyme. As a matter
of fact "the same reaction" may not be possible at all without an enzyme. Most enzyme processes represent a sequence of several reactions with several stable intermediates. It could, for instance, so happen, that this path of the overall reaction, cannot be realized in the absence of an enzyme. Fortunately, as a rule it is,
probably, not true. KOSOWER [6.10] introduced a very convenient concept of a congruent model system, i.e., of a chemical system in which the same overall reaction

with the same stable intermediates is realized. These intermediates may represent complexes of substrates, of substrate and ions (e.g., H^+), or of substrate and catalytic groups.

Below, statements concerning the degree of activity of enzymes and the reaction acceleration will invariably imply that the enzyme reaction is compared with the corresponding congruent reaction.

Reaction paths, determined by stable intermediates (which, in principle, can be isolated) coincide for the enzyme and the model congruent systems. As far as the reaction paths occurring in the course of individual reaction steps are concerned, they are, as a rule, known neither for the enzyme nor for the model system. These paths are, most definitely, different: otherwise catalytic effect would be impossible.

All the vast number of points of view published on the origin of the extremely high catalytic activity of enzymes may be subdivided into four groups according to the following statements:

1) Complex formation with an enzyme leads to an increase of concentration and to mutual specific orientation of reacting particles (substrate molecules, catalytic groups), to an increase of the number of effective collisions, and, consequently, to the increase of the frequency factor in the Arrhenius equation (of ΔS_a in the equation of activated state theory).

2) Enzyme-substrate complex formation lowers the reaction activation energy, ΔE_a. This is due to the changes in electron density distribution over the substrate molecule, to its induced deformation, etc.

3) In an enzyme-substrate complex, large thermal energy fluctuations on certain vibrational degrees of freedom, on certain bonds are more probable than in the individual substrate molecule. An enzyme gathers, as it were, the thermal energy, and directs it to a bond undergoing the chemical transformation.

4) The energy liberated during sorption of substrate on the enzyme (during substrate-enzyme complex formation), during one of the stages of a multistage enzyme reaction, or in the course of chemical transformation in one-step enzyme reaction, does not dissipate but is recuperated and used to overcome the activation barriers of subsequent steps or elementary acts.

Let us consider these variants of explanations in detail. The first two are the most natural and may be called classical. Indeed, if the reaction rate constant is represented in the form of a product of two factors (entropy and energy factors), and if this constant increases in an enzyme process, as compared with a noncatalyzed one, then at least one of the factors must necessarily be increased. Let us begin with the entropy factor.

6.2.1 Enzyme Increases the Activation Entropy

This type of explanation has been discussed in detail in many publications (see,
e.g. [6.11]). The meaning of the effects supposed to be taking place can be summed
up in the following way. In a congruent model system, the chemical transformation
requires the drawing together of substrate molecules, ions and substrates, catalytic
groups (if any) and substrates, and also the mutual relative orientation of reacting particles. According to the transition state theory, an elementary act of a reaction proceeds along its unique path with the lowest possible activation barrier.
Thus, the structure of the transition state must be totally determined. These effects (the drawing together, orientation, etc.) introduce, therefore, a considerable negative contribution to the activation entropy in a model system. If complex
formation with an enzyme leads to the necessary initial relative orientation of substrate molecular, of catalitic groups and substrate molecules, etc., the entropy
of the substrate-enzyme complex would have already been lowered. Therefore, the activation entropy increases, as well as the reaction rate. Increased reagent concentration due to complex formation does not always lead to reaction acceleration.
If there is an abundance of one of the reagents (for instance, water in hydrolytic
processes), then, due to the reaction being localized within rather few complexes
(enzyme concentration is always low), the process may proceed even slower. Orientation effects are, however, always favorable. The calculated value of reaction
acceleration due to these entropy effects may, in principle, be arbitrarily high,
because it is determined by the range of solid angles between the interacting particles, which one assumes to be appropriate for the normal proceeding of the reaction. The less this range is, the more precisely the enzyme must fix reagent molecules in space, and the higher will be the resulting acceleration of the reaction. KOSHLAND [6.11,12] estimated that such entropy effects may be as high as
18 orders of magnitude, and may, thus, make a major contribution to catalytic action
of the enzyme.

Several years ago MILSTEIN and COHEN [6.13] published an interesting paper, the
aim of which was to verify experimentally the possibility of changes in the reaction rate due to the fixation of the mutual arrangement of reacting groups (i.e.,
due to a considerable restriction of translational and rotational degrees of freedom). The authors studied the kinetics of acid-catalyzed ester formation. Five different processes (see diagrams on page 97) were compared.

From 1 to 5, the relative position and orientation of hydroxy- and carboxy-
groups, between which an ester bond is being formed, are gradually more and more
fixed. Reaction 1 is bimolecular, and the initial orientation of reagents is,
practically, unrestricted. In monomolecular reactions 2-5, the freedom of movement
of reacting groups is successively diminished due to the increasing steric hindrances. Table 6.1 presents the values of thermodynamic and kinetic reaction

1)
$$\text{phenol (OH)} + CH_3COOH \xrightarrow{H_3O^+} H_2O + \text{phenyl acetate}$$

2)
$$\xrightarrow{H_3O^+} H_2O +$$

3)
$$\xrightarrow{H_3O^+} H_2O +$$

4)
$$\xrightarrow{H_3O^+} H_2O +$$

5)
$$\xrightarrow{H_3O^+} H_2O +$$

Table 6.1

Reaction	K	ΔF^0 [kcal·mol^{-1}]	$k_{H_3O^+}$
1	$3,8 \cdot 10^{-6}$	7390	$\sim 10^{-10}$ M^{-1}.s^{-1}
2	0,04	1980	$5,94 \cdot 10^{-6}$ s^{-1}
3	0,62	290	$6,22 \cdot 10^{-6}$ s^{-1}
4	25,67	-1950	$2,62 \cdot 10^{-2}$ s^{-1}
5	> 99	<-2770	$5,9 \cdot 10^{5}$ s^{-1}

parameters: equilibrium constant, K, standard free energy of reaction, ΔF^0, and effective rate constant of acid-catalyzed reaction $k_{H_3O^+}$.

The transition from a bimolecular to a monomolecular reaction and the subsequent restriction of the freedom of movement of reacting groups, thus, ultimately lead

to the increase in reaction rate by more than 15 orders of magnitude (this statement holds true if reagent concentrations in a bimolecular reaction do not exceed
1 M).

At first sight it seems that these results indeed confirm the possibility of a
great acceleration of enzyme reaction owing to entropy factors, and may be regarded
as experimental proof of the first theoretical variant of enzymatic catalysis explanation. It is, however, quite easy to see that this is not true. Reaction 5, for
instance, that is to be compared with reaction 1, is actually the reaction of a
new, *already synthesized* compound. If we are speaking about an enzymatic, catalytic
process, it is necessary to consider all its stages from the formation of a substrate-enzyme complex having a given fixed mutual orientation of reagents to the
product desorption into the homogeneous phase, into solution. Evidently, the formation of a substrate-enzyme complex must, in this case, be accompanied with a decrease in the system entropy by at least the same value, as the value of activation
entropy increases for the reaction studied. In order to have an increase in reaction rate by 18 orders of magnitude required by KOSHLAND, the system entropy must
be diminished, due to substrate-enzyme complex formation, at least by $18.2.23 = 83$
e.u. If the formation of this complex is to be realized at room temperature, the
binding energy between the substrate and the enzyme cannot be less than
$83.300 \approx 24000$ cal.mol^{-1}.

What is the physical cause of reaction acceleration due to the "right" initial
orientation of reacting molecules? Evidently, at the same temperature, the number
of "successful" (i.e., those leading to the chemical transformation) collisions
in this case increases. As a matter of fact, the first variant of the explanation
is based on the assumption of all the postulates of classical chemical kinetics
of gaseous reactions being fulfilled. The relative movement of reacting molecules,
their movement along the reaction coordinate, must, therefore, be regarded as thermal movement. The existence of a compulsory (as we have seen) strong bond with
enzyme, which effectively freezes all translational and rotational degrees of freedom, must, however, influence the reaction activation energy. If we continue in the
assumption that, in a substrate-enzyme complex, it is the substrate molecule that
undergoes a chemical transformation, the strong bond between the substrate and the
enzyme must then hinder the movement along the reaction coordinate. Moreover, the
completion of a catalytic process requires that the bonds between enzyme and reaction products be cleavaged and the enzyme molecule must return to its initial
state. If the bond with the enzyme does not change in the course of the substrate
chemical transformation it can happen that the limiting stage of the overall process
will be the dissociation of the product-enzyme complex. But, if the energy of the
bond with the enzyme diminishes in the course of substrate-product transformation,
it means that the substrate chemical transformation is accompanied by an increase
of system energy. Therefore, at least a part of the energy liberated in the course
of the reaction must be transformed into a new form without dissipation. The

question arises as to what the mechanism of this transformation is. Strictly speaking, this question concerns the fourth variant of the explanation. Thus, in the realm of traditional classical chemical kinetics it is difficult to consider the statement "an enzyme increases activation entropy" as explaining the enzymatic catalysis. This increase in itself does not necessarily lead to the increase of reaction rate. Similar reasoning is found in [6.14]. Summing up, we can suggest that the specificity of substrate bonding by enzyme, leading to the unique mutual arrangement and orientation of reacting groups, is necessary for the enzymatic process, but cannot in itself explain the increase of enzymatic reaction rate as compared with the rate of the same reaction in a congruent model system.

6.2.2 Enzyme Lowers the Activation Energy

This variant explanation is, perhaps, the most popular. The reason lies in the fact that the measurements of temperature dependence of enzyme reaction rates usually give lower activation energy values than those for analogous nonenzymatic reactions (see, e.g. [6.14]). It was, however, shown in the preceding chapter (and will be verified in this chapter) that, for enzymatic reactions, the Arrhenius-like determinations of activation energies are meaningless. Therefore, the experimental data on activation energies cannot be, generally speaking, considered a foundation for the correctness of this variant of explanation.

In the final analysis, all the mechanisms of activation barrier lowering assumed in enzymatic catalysis come to two effects: a) rearrangement of electron density in the substrate molecule due to interaction with various enzyme groups (ion-dipole and dipole-dipole interactions, hydrogen bonds, etc.), and b) "concerted" action of various catalytic groups of enzymes, due to which a reaction proceeding without enzyme through several time-separated stages, has in the presence of enzyme an activated state of lower energy. Naturally, both these explanations remain mere words if specific interactions, the structure of the substrate-enzyme complex and relative displacements of its parts taking place while the system moves along the reaction coordinate have not been indicated.

Of course, electron density rearrangement in a substrate molecule due to the interaction with an enzyme does actually take place and changes the substrate reactivity. In this respect, the existing ideas concerning the mechanisms of enzyme action do not add anything to the words usually written in papers dealing with catalytic chemical reactions. Schemes of certain processes and corresponding reflections can be found in the reviews cited above.

Recently one other mechanism of activation energy lowering in enzymatic catalysis has been broadly discussed in the literature. We are referring to the so-called "rack" mechanism proposed by EYRING et al. as early as 1954 [6.15]. The essence of this mechanism is clear from Fig.6.1. Enzyme-substrate complex formation leads to substrate deformation, to the stretching of one of the bonds in the substrate molecule, and to the corresponding lowering of its break activation energy.

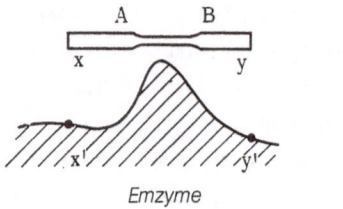
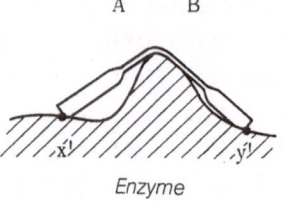

Emzyme   Enzyme

Fig.6.1.
The "rack" model

Let us discuss this mechanism in greater detail. The lowering of activation energy of the bond A-B dissociation is in this case, evidently, due to the increase in the bond A-B initial energy which, in its turn, is compensated by the energy of substrate sorption on the enzyme. After the reaction has been completed, i.e., after the rupture of the A-B bond, steric hindrances vanish and, consequently, the energy of reaction product sorption must become greater than the sorption energy of the initial substrate. To be catalytic, the overall process must include the enzyme return to the initial state, i.e., must include the desorption of reaction products. Bearing in mind the increase in binding energy between reaction products and enzyme that has just been mentioned, one may apprehend that the rise in the activation energy of product desorption (by value approximately equal to the de- crease of the A-B bond energy due to substrate bonding) will make desorption the limiting stage of the overall process. Therefore, without a detailed analysis of the mechanisms of all stages in each specific case, the "rack" mechanism cannot be regarded as an explanation of activation energy lowering in enzymatic catalysis.

The entropy, as well as energy, variants of the explanation of high rates of enzymatic reactions both assume the existence of a unique structural correspondence between the substrate and the enzyme. In current literature it is customary to explain on this basis the extremely high structural specificity of substrate-enzyme interaction. Formerly structural fitness was considered to be static ("lock and key" hypothesis). In 1958 KOSHLAND formulated the theory of induced fit [6.12,16], according to which structural fitness arises due to conformational changes in the enzyme molecule induced by the substrate. This theory was an important stage in the development of theoretical ideas in enzymology because it placed the emphasis on the compulsory nature of enzyme conformational changes in the course of the inter- action with the substrate, and these changes, as we have seen in Chap.5 and shall see in this chapter, always indeed take place. However, the induced fit theory does not affect our reasoning concerning the two first variants of enzymatic activity explanations. In the realm of this theory, the enzyme conformational change induced by the substrate comes to an end with the formation of a substrate-enzyme complex and does not bear any direct relation to the elementary act of substrate molecule chemical transformation[2].

---

[2] About the compulsory character of enzyme conformational transitions due to electron density changes induced by substrate, see also [6.17,18].

6.2.3 Enzyme Heightens the Probability of Useful Energy Fluctuation

This variant explanation was proposed by MOELWYN-HUGHES in 1950 [6.19]. It is founded on the well-known work of HINSHELWOOD concerning the theory of unimolecular reactions in gaseous phase at low pressures [6.20,21].

At high gas pressures an elementary act of chemical reaction, e.g., dissociation of the ith bond, takes place in molecules for which the average energy of the vibrational degree of freedom, corresponding to the stretching vibration of the ith bond, ε_i, is greater than or equal to the activation barrier of bond dissociation, E_a, i.e.,

$$\varepsilon_i \geq E_a \quad . \tag{6.1}$$

If the gas pressure is low (i.e., if the collision frequency is not high), the act of reaction may take place when a milder requirement is fulfilled

$$\sum_i \varepsilon_i \geq E_a \quad . \tag{6.2}$$

It means that all molecules whose total vibrational energy is greater than or equal to the activation barrier can enter into a reaction. Such a requirement is enough because, at low collision frequency, there is sufficient time for an "energized" molecule to realize a fluctuation of vibrational energy, due to which all this energy will be localized on one i^{th} degree of freedom. At a given temperature, the number of molecules satisfying (6.2) is, naturally, much greater than the number of molecules satisfying (6.1). At low enough pressures the fulfillment of (6.2) is equivalent to the lowering of measured activation energy E_a' as compared with activation barier energy E_a

$$E_a' = E_a - (S - 1)RT \quad , \tag{6.3}$$

where 2S is the number of quadratic terms (of kinetic and potential energies), necessary for unique determination of molecule energy.

The Hinshelwood-Kassel theory was discussed in detail and criticized in [6.22]. We shall not dwell at length here on this criticism. It is indeed clear that this theory is based on the assumption that the excess vibrational energy of a molecule may not have enough time to dissipate due to collisions with other molecules. It means that this theory holds true only for gaseous reactions at low pressures. In the case of enzymatic processes in condensed phase, the assumption on which this theory is based, is not definitely fulfilled.

6.2.4 Energy Recuperation

I think that the first scientist to discuss the possibility of processes which were later called "the processes of energy recuperation", was BAUER in his brilliant book "Theoretical Biology", published in 1935 [6.23]. According to BAUER, protein macromolecules exist within the cell in a specific deformed state. During

an enzymatic reaction proteins undergo transitions to the equilibrium state, and
the liberated energy is used to overcome the activation barrier. The energy liberated in the course of enzymatic reaction, in turn, does not dissipate but is used
to return the protein molecule to the initial deformed and strained state. The
boldness and depth of BAUER's hypothesis are surprising (it should be remembered,
that it was published as early as 1935 when nobody did actually know anything
about protein conformational changes). The subsequent authors have added but little
to BAUER's hypothesis (except for terminology).

The term "energy recuperation" was introduced by KOBOSEV in 1960 [6.24]. The
meaning of this term is connected with the physical interpretation of the so-called
aggravation effect [6.25], i.e., of nonspecific effects of the large catalyst molecule upon the reaction rate. According to KOBOSEV, the role played by the protein
macromolecule in an enzymatic reaction consists in the utilization of energy,
liberated during the reaction, for the activation of the catalytic center.

A more concrete physical mechanism of energy recuperation in the course of enzymatic catalysis was proposed by CHURGIN et al. [6.26]. These authors assumed that
the energy liberated during the reaction is stored in the form of elastic deformation
up to the beginning of the next catalytic act. Rather rough calculations carried
out by the authors and based on the assumption of a protein molecule behaving as an
elastic solid body showed that storage of sufficient quantity of energy (about
$10 \ kcal \cdot mol^{-1}$) and the spreading of an elastic deformation region through all the
globule (about 50 Å) are quite possible. Naturally, such calculations neither prove
nor disprove anything.

The most important achievements of this work were the separation of macromolecule degrees of freedom into the normal and the "constructional" (mechanical), and
the suggestion of the dissipation of energy localized on these specific degrees of
freedom being slow.

The same ideas were further developed by Chernavsky and Hurgin in their reports
to the 5th Intern. Symposium on Chemistry of Natural Compounds in June 1970 in
Riga. They concluded that, for one-step enzymatic catalysis (especially if the enzyme function is to lower the entropy portion of the activation barrier), energy
recuperation is not necessary. On the other hand, in the case of multistep enzymatic processes, energy storage and recuperation are necessary to ensure the step-by-
step changes of structure, i.e., to secure optimal structures during the transition
from one stage of the process to another. The sources of energy stored by an enzyme may be substrate sorption on the enzyme, or exoergic stages of the enzymatic
process itself.

SHNOLL [6.27] gave a very clear description of the general mechanism of enzymatic catalysis based on energy recuperation. He wrote: "The conformation of an isolated enzyme molecule differs from that of protein in the substrate-enzyme complex.
The protein macromolecule is constructed *in such a way* that after the realization

of the substrate transformation step (after random thermal fluctuation), the liber- ated energy transforms enzyme molecules into the specific nonequilibrium state of thermodynamically unfavorable conformation. For kinetic reasons (a large activation barrier) the macromolecule can remain in this state long enough (in general, this conformation may be, probably, characterized by an extremely great lifetime). The nonequilibrium macromolecule may be transferred into a new probable state only due to the contact with a substrate. We have a funny, but quite realistic paradox: the substrate catalyzes the enzyme transition into equilibrium state. In return, the substrate undergoes the necessary transformations, whose activation barrier di- minishes at the expense of the *conformational* energy of the enzyme macromolecule".

The merits of this variant enzymatic activity explanation are evident. Confor- mational changes prove to be compulsory at all stages of catalyzed transformations. These theoretical considerations have, however, two rather serious shortcomings.

First of all, the physical mechanism of the whole protein macromolecule tran- sition into a deformed state with higher conformational energy directly induced by a local microchemical act remains quite obscure. This variant explanation only states that "the protein molecule is constructed in such a way that...".

Secondly, the mechanism of the action of energy, stored in the form of elastic deformation of the whole conformationally changed macromolecule, is not easy to understand. The authors considered an enzyme to be a construction, a device having specific degrees of freedom, but did not ask what is, in this case, the meaning of activation energy, the value of which this stored energy must lower (see also [6.28,29]).

6.3 Conformational Changes as Necessary Steps of Enzymatic Processes

In Chap.5 many facts were presented indicative of conformational transformation of the whole protein macromolecule caused by local microchemical changes due, for in- stance, to the changes of the active centered charge (redox processes), to low- molecular compound addition, etc. Since substrate-enzyme complex formation is a necessary stage of any enzymatic process, we can presume that, in the course of enzymatic reaction, conformational changes of enzyme macromolecules are always re- alized. This fact can now be considered to have been experimentally proved. Several examples will be discussed below.

The results presented in Sect.5.6 concerning the protein conformational changes which accompany the functioning of "honorary enzymes", haemoglobin and cytochrome C, can be, naturally, also considered a corroboration of the above thesis.

KAJNE and SUELTER [6.30] studied the effects of complex formation with the sub- strate, of the binding of activating cations, and that of temperature upon the protein macromolecular state of a very important enzyme—pyruvate kinase. All these

factors actually brought about qualitatively the same changes. Conformational
changes were recorded by means of differential absorption spectra in the 280-300 nm
region. The observed changes in spectra indicated that considerable changes take
place in the environment of certain aromatic chromophores, e.g., tryptophan, viz.,
the transition from aqueous to hydrophobic environment. There cannot be any doubt
of the fact that substrate binding, addition of activating ions, and temperature
changes lead to similar conformational changes of pyruvate kinase globula.

Very interesting data concerning the conformational changes of the same enzyme
due to substrate binding were obtained by REUBEN and KAINE [6.31] using NMR on the
nuclei of thallium natural stable isotope Tl^{205} (nuclear spin 1/2). These authors
studied the complex of pyruvate kinase with Tl^+ and Mn^{2+} ions (both these ions are
enzyme activators). The recorded parameters of the NMR spectrum of Tl^{205} allow one
to estimate the distance between the Tl^+ ion and the paramagnetic Mn^{2+} ion, attached to different groups of the protein molecule. For free protein, this distance
proved to be 8.2 Å. Substrate (phosphophenylpyruvate) binding decreased this distance almost by half (to 4.9 Å). This proves the existence of essential protein
conformational changes accompanying the phosphophenylpyruvate binding.

Important results concerning the conformational transitions of aspartatetrans-
carbamylase (ATCA) were obtained by HAMMES and co-workers [6.32]. ATCA catalyzes
the formation of carbamyl-α-aspartate from α-aspartate and carbamylphosphate in
the first stage of pyrimidine biosynthesis. The enzyme molecule has a quaternary
structure and six catalytic subunits (with molecular weight 33000) and six regulatory subunits (with molecular weight 1700). Substrate molecules interact with
catalytic subunits only. Activator (ATP) and inhibitor (CTP, Br-CTP) molecules can
be attached only to the regulatory subunits ensuring the changes in the activity
of the catalytic subunits of this enzyme. The authors [6.32] studied the kinetics
of various transitions of the whole enzyme and its individual subunits between conformational state under the action of substrate, activator, and inhibitor molecules
in various combinations. The changes were recorded by means of difference UV absorption spectra, the methods of relaxation kinetics being applied. The authors
found conformational transformations of various duration: from hundreds of micro-
seconds to tenths of milliseconds. Fast conformational changes are associated with
catalytic processes, and slow ones with regulatory processes.

Interesting results were also obtained on one of the most widely distributed
enzymes - D-glyceraldehyde-3-phosphate dehydrogenase (GAPD), that catalyzes oxy-
genative phosphorylation of D-glyceraldehyde-3-phosphate into 1,3-diphosphoglyceric
acid with NAD^+ as coenzyme. The enzyme molecule contains four subunits. With the
help of spin label technique BALTHASAR [6.33] showed coenzyme binding to result in
protein conformational changes in the active center region (see also [6.34]).

Complex formation between GAPD and coenzyme NAD^+ was also studied by KIRSCHER
and co-workers [6.35,36]. They recorded differential absorbtion spectra (between

3000 and 5000 Å) of enzyme solutions in the presence of NAD^+ at different temperatures, as well as the kinetics of the transition between states after a sudden increase of sample temperature (temperature jump method with dead time of about $2 \cdot 10^{-6}$ s). It is noteworthy that the spectrum sharply depends on temperature in the region of 3-40°C without any isobestic point. This may imply gradual changes of the chromophore environment with temperature, due to the conformational rearrangement of the protein globule. In fact, with the help of the temperature jump method a fast monomolecular relaxation process ($\tau < 10^{-5}$ s) associated with the rearrangement of the active center chromophore environment, rather than with the shift of equilibrium between enzyme and coenzyme, was revealed. The enzyme conformational relaxation caused by coenzyme dissociation or binding after sudden temperature changes proceeds much slower (τ from 0.1 to 3.0 s, depending on coenzyme concentration). With the help of the stop-flow method the authors established that complex formation with NAD^+ proceeds in two steps: a fast bimolecular step ($\kappa \approx 10^5$ $mol^{-1}s^{-1}$) is followed by a slow monomolecular relaxation step ($\kappa \approx 0.3$ s^{-1}).

Essential results have been recently obtained with the help of spin and fluorescence label technique. There are several good reviews [6.37-39]. The principles of spin label technique, the nature of information obtained by the measurements of label ESR signal parameter changes, caused by the changes of label environment, can be found in special publications [6.40-46]. Interesting data were obtained in [6.47,48]. In the case of glutamataspartatetransaminase, the action of specific substrates and inhibitors on the active center leads to conformational changes recorded by changes in the ESR spectrum of label 16-17 Å away from the active center. The same results were obtained in the investigation of the action of specific inhibitors on lysozym—a comparatively simple enzyme without a quaternary structure. The conformational changes of myosin during ATP enzymatic hydrolysis were also studied [6.49-54]. With the help of a spin label attached to the S-1 thiol group of protein at a distance of 6 Å from the active center it was found that in the course of enzymatic hydrolysis the conformational state of myosin molecules differs from that for free myosin and for the myosin equilibrium complex with reaction product ADP.

6.4 The Effect of Temperature on Enzymes and on the Activation Parameters of Enzyme Reactions

All the results discussed in Chap.5 concerning the temperature dependence of protein conformation, the inapplicability of the Arrhenius equation to protein reactions, the absence of physical meaning in the activation parameters of these processes, the so-called compensation effect, retain their validity for enzymatic reactions. Certain experimental data in Chap.5 on the temperature dependence of the

heat effects of protein reactions with low-molecular compounds and on the gradual
structural rearrangement of protein and the changes in the reaction path with
temperature, directly concern the interaction of enzymes (chymotrypsin, aldolase,
etc.) with inhibitors. In this section the data on the activation parameters of
enzymatic reactions and their temperature, pH, ionic strength dependences will be
discussed in greater detail.

A general analysis of the activation parameters of enzymatic reactions can be
found in [6.55]. Subtle changes of enzymes caused, for instance, by the changes in
the solvent salt content, or by one isozyme being replaced by another, can lead
to sharp shifts in ΔE_a and ΔS_a values (according to Arrhenius), this not being
accompanied, however, with essential changes in the reaction rate. For instance,
in [6.56] lactate dehydrogenase (LDH) from frog heart muscle is represented by three
isozymes: AAAA, BBAA and BBBB (enzyme molecule with a quaternary structure has four
subunits of two different types A and B). LDH from frog skeleton muscle contains
only one isozyme AAAA. The rates of reactions catalyzed by LDH's of heart and skeleton muscle are almost equal (the difference does not exceed one order of magnitude,
the activation free energy values ΔF_a being 22.8 and 24.1 kcal·mol^{-1}, respectively).
The values of activation energy, ΔE_a, for these enzymes were, however, found to be
59.1 and 29.5 kcal·mol^{-1}, and those of activation entropy, ΔS_a, 109 and 15.9 e.u.,
respectively, when one passes from heart to skeleton LDH, the exponential factor
in the Arrhenius rate constant equation increases ~10^{21} times, and the pre-exponential factor decreases ~10^{20} times. The possible physical mechanisms responsible
for such great changes of the activation parameters of similar processes, which do
not differ essentially in their sole measurable characteristic, the reaction rate,
are difficult to imagine.

Great changes in activation parameters without any significant alterations in
the process rate for the same enzyme are often observed in the case of substrate
replacements. For instance, the bimolecular rate constant of enzyme-substrate complex formation for chymotrypsin is 6.6 mol^{-1}s^{-1} (at pH 7.8,25°C, with methylhydro-
cinnamate as substrate). After the replacement of this substrate by methyl-d-
β-phenyllactate at the same conditions the rate constant becomes 4.0 mol^{-1}s^{-1},
i.e., decreases by the factor of 1.6 only. At the same time, the pre-exponential
factor in the Arrhenius equation decreases, and the exponential factor increases
by 5-6 orders of magnitude (see [6.57]). There are many of such examples in the
literature.

Especially great attention was given in the literature to the temperature dependence of activation parameters. The situation discussed most often is the one
where activation parameters undergo sudden changes (breaks on Arrhenius anamorphoses)
at certain temperatures. The corresponding data can be found in many reviews and
original monographs (see, e.g. [6.38,55,58,59]). In these cases the effects are
essentially the same as those noted above: there are but small changes in reac-

tion rate, and great changes in activation parameters measured with the help of the Arrhenius equation. Three types of interpretations of these breaks have been given in the literature. According to the first type of interpretation, there are really no sudden changes of true activation parameters, and enzymatic reactions obey an Arrhenius equation with one true activation energy (see, e.g., [6.57]). The rather gradual changes of activation parameters may, however, occur due to the fact that we do not measure the true rate constant of the elementary step of reaction, but the summary reaction rate, determined by various elementary equilibriums and rates. This may lead to most peculiar temperature dependences of observed activation parameters. This type of explanation can only be seldom found in current literature: too many studies have been carried out where the temperature dependence of the individual steps of enzymatic processes were measured, and strict proof of the existence of sharp changes in activation parameters in the interval of 2-3°C have been found. As an example we can cite the work of TALSKY and MÜLLER [6.60] who succeeded in measuring in detail (with intervals of about 0.5°C) the temperature dependence of cytidin-2',3'-monophosphate hydrolysis by ribonuclease and of the substrate-enzyme complex formation constant as well. They found that the breaks on Arrhenius anamorphoses are in reality the continuous transitions of complicated shape which occupy 1.5-3°C.

According to the opinion that was widespread some time ago, the breaks on Arrhenius anamorphosis indicate that the reaction under study proceeds through two consecutive stages with different activation energies, each stage determining the measured reaction rate within the corresponding temperature interval above or below the transition point (see, e.g. [6.61]). A sufficiently sharp transition can be recorded only if activation energies of the two stages differ by more than 16 kcal·mol^{-1} [6.62,63], and this is usually not realized. Besides, in this case, the stage with a higher activation energy must determine the process velocity at low temperatures and the stage with a lower activation energy that at high temperatures. It is, indeed, often true. For instance, enzyme hypoxanthine phosphoribosyltransferase, which catalyzes the transfer of phosphoribosyl residue from 5-phosphoibosylpyrophosphate to hypoxanthine, gives a sharp break on the Arrhenius anamorphosis at 23°C [6.64]. At lower temperatures E_a = 51.0 kcal·mol^{-1}, and at higher temperatures E_a = 13.1 kcal·mol^{-1}. This corresponds to the spasmodic increase of the exponential cofactor (and to the decrease of the pre-exponential cofactor) by about 28 orders of magnitude. At the same time, one can find quite the opposite situations. For instance, a very similar enzyme, the inosine monophosphate pyrophosphate phosphoribosyltransferase from yeasts [6.65], gives a break on the Arrhenius anamorphosis at 19°C. In this case, however, at lower temperatures ΔE_a = 5.7 kcal·mol^{-1}, and at temperatures higher than 19°C ΔE_a = 11.3 kcal·mol^{-1}, which corresponds to the decrease of exponential and to the increase of pre-exponential factors by more than 4 orders of magnitude.

The most popular at present is the third hypothesis: temperature dependence of activation parameters is caused by enzyme macromolecule conformational transition. The enzyme can exist in two forms having different properties, including ΔE_a values of the catalyzed ensymatic reaction. MASSEY et al. [6.66] found that, for d-amino acid oxidase, Arrhenius anamorphosis has a break at $14^\circ C$ (in the case of d-alamine oxidation): at higher temperatures $\Delta E_a = 10.3$ kcal·mol^{-1}, and at lower temperatures $\Delta E_a = 16.9$ kcal·mol^{-1}. In the same temperature interval (between 10 and $20^\circ C$), the authors observed changes in sedimentation constant, optical density in UV spectral region, and tryptophane fluorescence intensity. The changes in UV absorption are of continuous nature, and ΔD is a linear function of temperature without breaks. At the same time, the temperature dependence of tryptophane fluorescence intensity may indeed be interpreted as transition of molecules between two conformations one of which is stable at temperatures lower than $10^\circ C$, and the other at temperatures higher than $20^\circ C$. According to the authors, transition enthalpy $\Delta H \approx 78$ kcal.mol^{-1} Arrhenius anamorphosis for d-methionine (instead of alanine) has a break at $24^\circ C$ (instead of $14^\circ C$). These data leave no doubt as to the fact that, in the temperature region investigated, certain structural changes of the enzyme take place. There is, however, no proof that the transition occurs at just the temperature corresponding to the break on the Arrhenius anamorphosis. Moreover, it seems that the "transition point" depends on the registration method.

LEHRER and BARKER [6.67] gave the same interpretation to their kinetic study of fructosediphosphate splitting by rabbit muscle aldolase. They measured the temperature dependence of thermodynamic parameters of the overall reaction (using reaction equilibrium constant), that of substrate-enzyme complex formation (using Michaelis constant) and that of activation parameters (using V_{max}). The best

Table 6.2

	Overall reaction		Complex formation		Activation	
	Temperatures					
	lower ($21.2^\circ C$)	higher ($30^\circ C$)	lower ($20^\circ C$)	higher ($31.9^\circ C$)	lower ($20^\circ C$)	higher ($31.9^\circ C$)
	than the temperature of conformational transition					
ΔF[kcal·mol^{-1}]]	6,5	6,4	-7,2	- 7,4	15,4	15,5
ΔH[kcal·mol^{-1}]	14,30	10,6	11,5	-13,8	18,2	13,4
ΔS[e.u.]	26,5	13,8	63,8	-20,9	9,2	-7,3

approximation of Arrhenius anamorphosis is represented by two straight lines intersecting at $25^\circ C$. This temperature has been regarded as the temperature of transition between two "conformers". Anamorphosis satisfies the experimental data equally well if we accept for transition enthalpy any value between 10 and 1000 kcal·mol^{-1}. Table 6.2 presents experimental values of free energy, ΔF, enthalpy,

ΔH, and entropy, ΔS of processes studied at temperatures below and above the conformational transition temperature.

It can be seen that, judging by the values of parameters which have been actually measured, i.e., by transformation of activation free energy values, there is no transition at all. The difference between the conformers exists only due to the application of the Vant' Hoff or the Arrhenius equation.

Sometimes two conformers are insufficient to explain the experimental data. For instance, Arrhenius anamorphosis for the reaction of ATP and glucose-1-phosphate synthesis from ADP and glucose (enzyme-pyrophosphorylase from Rhodospirillum rubrum) has, in the presence as well in the absence of an activator (pyruvate), a rather complicated shape [6.68]. In the absence of pyruvate, the temperature dependence curve can be linearized in Arrhenius coordinates for temperatures lower than $26^{\circ}C$. In this region, the measured activation enthalpy equals 14.2 kcal·mol^{-1}. At higher temperatures, anamorphosis is no longer linear, the measured activation enthalpy drops rapidly, being equal to zero at $32^{\circ}C$ and becoming negative for temperatures higher than $38^{\circ}C$. In the presence of pyruvate, activation enthalpy is constant (11.2 kcal·mol^{-1}) up to $32^{\circ}C$ and reaches the zero value only at $52^{\circ}C$. These results can be explained if you postulate the existence of three conformers: a "high-temperature" and two "low-temperature" ones, with reversible transitions between them.

Temperature dependence of ESR spectra of spin labelled myosin and its ADP complex in the temperature interval from -2 to $+37^{\circ}C$ was studied in our laboratory [6.53]. We measured the correlation time (τ_{cor}) of a spin label covalently bounded by myosin SH-groups. It was found that, as a result of ADP binding, the protein structure in the vicinity of the spin label becomes looser. The character of τ_{cor} changes with temperature allows for two kinds of explanations.

1) Temperature changes lead to more or less continuous structural changes of the protein matrix. Each temperature has a corresponding different average myosin molecular conformation. 2) There are several quite definite protein conformations in solution at every temperature. Temperature changes lead to the shift of equilibrium between these conformations. In this case, the existence of at least five structural forms of enzyme in solution at equilibrium must be postulated to describe the τ_{cor} changes from -2 to $+37^{\circ}C$. The first explanation seems to be more verisimilar.

The results described in this and the preceding chapter imply that temperature anomalies of enzymatic reactions are caused by incorrect application of Vant' Hoff and Arrhenius equations to experimental data. At temperatures much lower than the disordering, denaturation temperature of a protein molecule, any temperature changes lead to a more or less pronounced rearrangement of the protein globule. It seems that this rearrangement usually concerns the configuration but not the conformation of the globule (see Sect.4.4). With temperature increase, the globule is distorted like a wire construction. In Sect.5.6 we saw what would be, in this case, the con-

sequences of treating protein reactions with the conventional methods of gas kinetics. All that was said in Sect.5.6 can be fully applied to enzymatic processes.

A natural question arises: is this statement about the gradual protein configuration changes with temperature consistent with numerous experimental data show-
ing sudden temperature changes of various physical properties of the systems investigated (label or intrinsic amino acid fluorescence intensity, correlation time
of spin labels, etc.). This contradiction is imaginary. In all cases, the existence
of such jump-like changes is connected with changes in the interaction between a
low-molecular indicator and the neighboring groups of protein molecules. The dis-
position of protein groups gradually changes with temperature, but the dimensions
of the low-molecular indicator remain practically constant. The distance between
the indicator and the protein groups changes gradually, but, due to the exponential
character of short-range intermolecular forces, the interaction changes abruptly
when the protein is passing through a configuration at a certain temperature. Thus,
the jumps observed unequivocally indicate that, within the temperature interval
studied, some configurational (and, may be, continuous) changes of protein globule
do occur, but the exact position of the "breaks" reflects the registration method
rather than the protein properties.

6.5 The Physics of Elementary Steps of Enzyme Catalysis [6.69,72,79,80]

Let us now sum up what has been said above. Analysis of the data given in the previous two chapters leads us to the following conclusion: during an enzymatic reaction the enzyme macromolecule undergoes constructional changes — conformational
transformations compulsory for its functioning. We have seen that activation parameters of enzymatic reactions, measured, or rather calculated, using the equations
of Arrhenius and of the absolute rate theory (even in relation to the individual
steps of a multistage process), do not have the same meaning that these quantities
have in the classical gas reactions, and, as a rule, do not have any physical
meaning whatsoever. When analyzing the kinetic aspects of enzymatic processes we,
nevertheless, postulated, explicitly or tacitly, that some true activation parameters, ΔE_a and ΔS_a, of enzymatic reaction exist, but that, unfortunately, we were
unable to measure them. What are the grounds for this postulate? We have seen that
the formal validity of the Arrhenius equation for reaction rate temperature dependence cannot be regarded as a foundation of any theory. The assumed existence of
true ΔE_a and ΔS_a seems to be based only on the habit of treating the results of
kinetic experiments using equations of the Arrhenius type.

This assumption is equivalent to the following picture of a chemical transformation process. There is a definite reaction path for any given temperature. This
path corresponds to the given (lowest among those possible) activation barrier,

ΔE_a. The reaction rate is determined by the number of molecules with energies sufficiently high to overcome this barrier. The changes of reaction rate with temperature are due to the following reasons: 1) changes of the activation barrier [in accordance with condition (5.3) or without it], and 2) increase of the number of high energy molecules with temperature.

These two reasons are related to different processes. The change of activation barrier is connected with conformational and configurational macromolecular transformations, and the increase of the number of high energy molecules (according to the equilibrium distribution) is related to the chemical transformation of the substrate. Thus, the postulate of the existence of true activation energy and entropy is based on the differentiation between the chemical reaction of a substrate and the conformational transformation of a substrate-enzyme complex. In some cases, this differentiation and, consequently, the whole picture may be justified even for protein reactions. However, as a rule, in the light of all the material presented in this chapter, this differentiation becomes meaningless. The chemical transformation of the substrate, accompanied by the rupturing of some bonds, or by the formation of new bonds, or by both, takes place simultaneously with the conformational transformation of the whole substrate-enzyme complex.

Let us suggest at this point a new concept of the elementary act in enzymatic catalysis. This concept is based on the postulate, according to which the conformational change of the substrate-enzyme complex, accompanying the substrate binding by the enzyme active center, is of the nature of relaxation and includes not only the rupture of the old and the formation of new secondary bonds in the protein macromolecule, but also chemical changes necessary to transform the substrate molecule into a molecule or molecules of a product. *This conformational relaxation is, essentially, an elementary act of enzymatic reaction, and the rate of substrate-product transformation is determined by the rate of this conformational change.*

Before analyzing the general scheme of an enzymatic process on the basis of this concept, let us clarify the meaning of the suggested relaxational nature of conformational changes. One can picture the following series of incidents after the "sudden" local change in the active center (charge change, substrate or inhibitor binding, etc.). The most pronounced changes in the electronic structure and the geometry of the active center occur during vibrational relaxation ($\tau \approx 10^{-12} - 10^{-13}$ s). It is, probably, this very process which accounts for the main optical density changes observed after photodissociation of oxy- and carboxyhemoglobin by laser flashes [6.70] (see 5.4). The delay of these changes relative to the flashes was not recorded with a resolution of $\sim 10^{-9}$ s. As has already been stated, these practically instantaneous changes concern only the substrate molecules and those atomic groups in the substrate-enzyme complex which are in direct contact with the substrate. Thus, after the relaxation of vibrational degrees of freedom of the active center, the following situation is realized. The substrate, active center and the nearest surrounding parts of the protein molecule are in a new local equilibrium state, but the other parts of the macromolecule remain unchanged. There is steric strain. Relaxation of the whole macromolecule to the new conformation must proceed in two stages. At first the changes that must be realized are those called above configurational changes which do not require any modification of the secondary bond net. These changes will, mainly, affect the geometry of the immediate sur- rounding of the active center, but not the whole molecular conformation. They re- quire coordinated motion of several atomic groups and are, therefore, much slower than the primary changes of the active center. As a rule, the changes in macromo- lecule parts in the vicinity of the active center lead to the changes of its elec- tronic characteristics. Additional small optical density changes in the heme spec- trum of hemoglobin, occurring within 10^{-7} s after photodissociation [6.70], probably reflect these secondary processes.

The remaining macromolecular changes are associated with relaxation to the new conformation of protein, that becomes kinetically available, due to the completion of local microchemical changes. According to the formulated postulate a catalyzed transformation is realized in the course of this relaxation of the substrate-enzyme complex the duration of which may be as great as milliseconds and more (see data and references in Sect.5.4)[3].

Proceeding from this postulate, let us consider a scheme of an enzymatic pro- cess [6.69,72]. We assume the catalyzed reaction to consist in the rupturing of one of the bonds in a molecule AB consisting of two atomic groups A and B. As a result of the reaction, the groups become individual molecules

$$A - B \rightarrow A + B \ . \tag{6.4}$$

Let us then assume that for one molecule AB the reaction (6.4) under fixed con- ditions (temperature, medium composition, pH, etc.) is accompanied by a decrease Δu_0 in the system's total energy. Reaction (6.4) may include not only the rupturing of the A-B bond but also, for instance, the addition of a proton and a hydroxyl ion to groups A and B (hydrolysis reaction). The corresponding energy changes, as well as those due to solvation, contribute to Δu_0. In considering the energetics of various steps of the enzymatic reaction we shall be dealing only with energy changes that are accompanying the transformation of one molecule (one macromole- cular complex). Therefore, we are always speaking about the total energy of the system, but not about its free energy (see also Chap.8).

Let us now conduct this process catalytically involving the specific enzyme, E. The reaction can be, naturally, divided into four steps (Fig.6.2).

a) Formation of the specific substrate-enzyme complex

$$A - B + E \rightleftharpoons (A - B)E \ . \tag{a}$$

---

[3]About the rearrangement times of proteins see also [6.71].

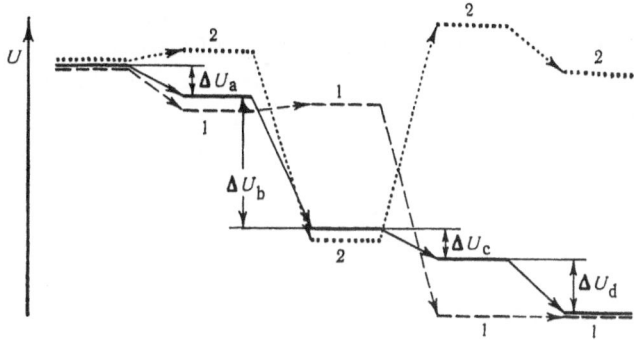

$$A-B+E \xrightarrow{(a)} (A-B)E \xrightarrow{(b)} A\widetilde{E}B \xrightarrow{(c)} \widetilde{E}+A+B \xrightarrow{(d)} E+A+B$$

Fig.6.2. Scheme of energy changes at different steps of enzymatic catalysis. (———) total system: substrate + enzyme; (----) subsystem 1: substrate + active center; (....) subsystem 2: remaining part of the enzyme. Initial energy levels of all systems are arbitrarily made to coincide

This fast step is a normal chemical reaction. The fact that one of the reactants, E, is a macromolecule, does not affect this step. We may consider the quasi-equilibrium state of step (a) to set in with unchanged configuration of the larger protein part. In this sense, the first step proceeds nonadiabatically[4].

Conformation E is now out of equilibrium, because there arises a new kinetically available conformational state of the substrate-enzyme complex with lower energy. A small part of complex (A-B)E near the active center, the electronic structure and geometry of which have undergone small changes, is, thus, at quasi-equilibrium, and the equilibrium of the main part of (A-B)E, the conformational relaxation of which is slow, becomes disturbed. Assume that, due to step (a), the total energy of the system decreases by Δu_a. This value can be formally subdivided into two parts

$$\Delta u_a = \Delta u_a^{(1)} + \Delta u_a^{(2)} \tag{6.5}$$

where $\Delta u_a^{(1)}$ is the energy change in the rapidly relaxing part of the substrate-enzyme complex, and $\Delta u_a^{(2)}$ that of the main part of the macromolecule, which had no

---

[4] Here and further the terms "adiabatic" and "nonadiabatic" have the same meaning as in mechanics. Let us imagine a system that can be divided into two subsystems in such a way that the overall process proceeding within this system can also be divided into two processes, one of which (process 1) represents the perturbation that initiates process 2. Let the characteristic times of processes 1 and 2 be τ_1 and τ_2, respectively. Process 1 is then called nonadiabatic if $\tau_1 < \tau_2$ and adiabatic if $\tau_1 \geq \tau_2$.

time to undergo conformational transformation. Evidently, $\Delta u_a^{(2)} > 0$ because the unchanged conformation has become unequilibrated, and constraints must develop in the regions close to the already changed active center. It has been assumed that, as a result of step (a), the total energy of the system decreases ($\Delta u_a < 0$). Therefore,

$$|\Delta u_a^{(1)}| > |\Delta u_a| \quad \text{and} \quad \Delta u_a^{(1)} < 0 \quad .$$

b) Slow relaxation of the substrate-enzyme complex to a new equilibrium and the substrate-product transformation

$$(A - B)E \rightarrow A\widetilde{E}B \tag{b}$$

where $\widetilde{E}$ is the new conformational state of the enzyme molecule. Chemical transformation of the substrate is realized as part of the conformational change of the macromolecular complex. During relaxation, the system energy decreases: $\Delta u_b < 0$. Quite formally we can again distinguish between $\Delta u_b^{(1)}$ (substrate and adjacent regions of the active center) and $\Delta u_b^{(2)}$ (remaining part of the macromolecule). The sign of $\Delta u_b^{(1)}$ depends on the nature of the substrate chemical transformation and on the active center structure. Let $\Delta u_b^{(1)} > 0$. Then, $\Delta u_b^2 < 0$ and $|\Delta u_b^{(2)}| > |\Delta u_b|$. Chemical transformation is realized by step (b). Therefore, this very step may be regarded as an elementary act of enzymatic reaction.

c) Decomposition of the product-enzyme complex

$$A\widetilde{E}B \rightleftharpoons A + B + \widetilde{E} \quad . \tag{c}$$

Like step (a), this step is a usual chemical transformation with rapid changes of the active center and with solvation changes of dissociating reaction products. The enzyme remains in conformation $\widetilde{E}$ which becomes unequilibrated and strained. Let $\Delta u_c < 0$. Obviously, $\Delta u_c^{(1)} < 0$, $\Delta u_c^{(2)} > 0$, $|\Delta u_c^{(1)}| > |\Delta u_c^{(2)}|$. As a result of the completion of steps (a), (b) and (c), the enzyme molecule becomes free again, but conformationally constrained. Therefore

$$\Delta u_a^{(2)} + \Delta u_b^{(2)} + \Delta u_c^{(2)} > 0 \quad . \tag{6.6}$$

This is shown in Fig.6.2.

d) Slow relaxation of the enzyme to the initial equilibrium state of a free macromolecule

$$\widetilde{E} \rightarrow E \tag{d}$$

$\Delta u_d = \Delta u_d^{(2)} < 0$, $\Delta u_d^{(1)} = 0$. Evidently

$$\Delta u_a + \Delta u_b + \Delta u_c + \Delta u_d = \Delta u_0 \quad . \tag{6.7}$$

The fast chemical steps (a) and (c) proceeding nonadiabatically relative to slow conformational changes of the macromolecule may be accompanied by a small increase (instead of a decrease) in the system energy. This will influence the quasi-equilibrium concentrations of complexes (A-B)E and A$\tilde{E}$B (in the case of statistical treatment of a catalytic process in the macroscopic volume). The rate of the whole process, however, seems to be limited, in most cases, by the slow relaxation steps (b) and (d). More complicated cases may be realized in enzymatic processes, when the first relaxation that does not involve chemical transformations is followed by the addition of a second substrate or of a cofactor molecule resulting in new conformational changes involving chemical transformations. The corresponding modification of the scheme is easy[5].

Let us now consider the kinetic aspects of the problem. According to the ideas presented in this book, the rate of chemical transformation of the substrate coincides with the rate of conformational relaxation of the substrate-enzyme complex [step (b)], i.e., with the velocity of system motion along the path determined by the specific mechanical degrees of freedom that are a function of macromolecular construction[6]. During relaxation, the energy of the whole system continuously decreases. One can, of course, formally distinguish atom shifts directly responsible for the changes in substrate molecular configuration, for its chemical transformation, and maintain that simultaneous changes in other parts of the macromolecule lead to a decrease of the effective activation barrier of this transformation. The very existence of specific mechanical degrees of freedom allows us to consider the shifts occurring at remote parts of the macromolecule as representing one elementary act. There is no sense in using the concepts of activation energy and activation entropy, in the same way as they are used in the activated state theory. As was shown above, condition (5.43) does not hold true for protein reactions. Within comparatively narrow temperature limits, where the kinetic studies of enzymatic processes are usually carried out, the temperature dependence of the reaction rate may be approximated satisfactorily enough by an exponential function. We have, however, already seen what errors may result from the formal application of the Arrhenius equation to the calculations of the activation parameters of enzymatic reactions. Strictly speaking, there are no activation parameters in the usual meaning of these words. The rates of relaxation steps change with temperature, due to changes in the initial macromolecular configuration and, consequently, in the path of its subsequent relaxation, rather than to those in the number of molecules with energies sufficient to overcome an activation barrier[7]. In addition to this, the

---

[5] About the electron-conformation interaction and its possible role in enzymatic catalysis see also [6.3,17,81].

[6] It does not exclude the possibility of a situation where the rate of free reaction product formation is determined by step (d). The above scheme is, naturally, idealized. If the product binding constant with $\tilde{E}$ is higher than that with E, the rate of unbound product appearance may be determined by relaxation step (d).

[7] Of course, the relaxation rate must itself be temperature dependent, but this dependence, in most cases, does not seem to affect the observed rate.

change of initial macromolecular conformation may lead to the decrease in the number of substrate molecules which are able, at step (a), to form a "proper" substrate-enzyme complex, i.e., a complex whose conformational relaxation is accompanied by the chemical substrate-product transformation. Probably, in most cases, the competing inhibitor differs from the substrate, since, due to small differences in binding and in paths of conformational relaxation, the latter is not accompanied by the rupture of bonds in inhibitor molecules. We can, therefore, say that, far from the temperature optimum, some substrate molecules behave like molecules of a competing inhibitor. In any case, the temperature dependence of the enzymatic reaction rate on both sides of the temperature optimum is to be explained in one and the same way: by structural changes of protein globula with temperature. Certainly, cases when the chemical steps, (a) and (c), rather than the relaxation steps, (b) and (d), are rate limiting cannot be excluded. Such cases, however, are probably an exeption.

The aforesaid can be verified experimentally. Conformational and configurational changes in proteins under the action of such unspecific agents, as temperature and pH may proceed much slower than the specific changes induced by the substrate. If this is actually the case, a situation can be realized when, for instance, the rate of an enzymatic process at the "new" temperature (pH) would correspond to the "old" temperature (pH). Indeed, these effects were found for enzymatic hydrolysis of ATP by myosin and of urea by urease [6.73]. It was shown in these experiments that not only does the configurational state change with temperature or pH but so does the path of conformational relaxation of the enzyme in the course of catalysis. These results provide evidence that the temperature dependence of the processes studied is determined by the changes in the initial state and in the path of conformational relaxation of the substrate-enzyme complex, rather than by the number of molecules able to overcome the activation barrier (see also [6.82]).

Furthermore, there is another reason for the importance of relaxation steps. The energy released in the course of fast chemical processes, as a rule, dissipates. Only slow relaxation steps may be utilized for useful work during muscle contraction or for ensuring the endothermic processes (see Chap.8).

The new relaxation concept of enzymatic catalysis presented in this chapter is just one of the possible variants of a more general scheme: protein-machine. This scheme also allows a more traditional approach [6.74], according to which the kinetically out of equilibrium state of the system after substrate-enzyme complex formation and the energy localization on specific degrees of freedom are used to lower the reaction activation barrier or to increase the effective collision number [6.83].

The relaxation concept and the traditional approach have one common feature: the mechanisms of direct and back reactions must be essentially different. It means that both these schemes can function only far from the state of thermodynamic equilibrium between substrate and product. Otherwise, the enzyme would change the reaction equilibrium constant, and this contradicts the laws of equilibrium thermodynamics.

From this point of view, the relaxation scheme was critically discussed in [6.75]. It is quite clear that the same type of reasoning could be applied to any scheme based on the concept "protein-machine", the "rack" scheme [6.15] included.

In all cases, when an enzyme is functioning as a mechanical device, the reaction mechanisms far from the state of thermodynamic equilibrium and close to it must be different.

Let us consider from this point of view the relaxation concept presented in this chapter. According to what has been stated above, enzymatic catalysis of chemical transformation

$$S \rightleftharpoons P \tag{6.8}$$

where S stands for the substrate, and P for the product, is realized far from thermodynamic equilibrium (6.8) as

$$S + E \underset{k_{-a}}{\overset{k_a}{\rightleftharpoons}} SE \overset{k_b}{\longrightarrow} P\widehat{E} \underset{k_{-c}}{\overset{k_c}{\rightleftharpoons}} P + \widehat{E} \overset{k_d}{\longrightarrow} P + E \tag{6.9}$$

$$P + E \underset{k'_{-a}}{\overset{k_a}{\rightleftharpoons}} PE \overset{k'_b}{\longrightarrow} S\widehat{E} \underset{k'_{-c}}{\overset{k'_c}{\rightleftharpoons}} S + \widehat{E} \overset{k'_d}{\longrightarrow} S + E \quad . \tag{6.10}$$

The direct reaction is described by (6.9), and the back reaction by (6.10). As was already stressed above, the elementary acts of fast steps (a), (c), (a') and (c') proceed and the corresponding quasi-equilibriums are reached with unchanged protein conformations. The relaxational stages can be described in two ways. The transition of the system to a new conformational state after substrate attachment to (or product dissociation from) the active center can be regarded as mechanical motion along one specific degree of freedom under the action of force, arising on the boundary between the relaxed and the unrelaxed regions of substrate-enzyme complex. This motion includes substrate-product transformation (or product-substrate transformation in the case of back reaction). The rate constant of such a stage, for instance, k_b, is not the true rate constant of a certain elementary chemical reaction. Its value depends on substrate concentration, on substrate-product ratio [i.e., on the degree of deviation from chemical equilibrium (6.8)]. Indeed, due to the slowness of relaxation (b), elementary acts of steps (a) are realized during the relaxation. In reality, relaxation (b) proceeds under the action of successive pushes rather than that of a permanent force. If the substrate concentration is very small, the pushes are rare, during the interval between them the enzyme has enough time to return to its initial conformation, and the resulting rate of relaxation will be diminished. The rise in product concentration, i.e., approach to the chemical equilibrium, acts in the same way. Thus, the "irreversibility" of step (b) (as well as of other relaxation steps) is due to the fact that system S-P is far from chemical equilibrium. Cyclic conformational changes of enzyme ($E \rightarrow \tilde{E} \rightarrow E$) are provided for by energy of $S \rightarrow P$ transformation far from chemical equilibrium.

The same process of conformational relaxation can be described in another way, namely, as a sequence of many elementary steps (turns around single bonds, changes in bond angles, rupture and formation of secondary bonds, etc.). Each of these steps is reversible, but the process as a whole is irreversible, if S-P system is out of chemical equilibrium. The path of overall relaxation is completely dependent on system construction. These two approaches are in reality indentical and lead to the same implications. In terms of the second description, relaxation step (b) can be represented as

$$S + E \rightleftharpoons \underset{a}{SE} \rightleftharpoons \underset{b^{(1)}}{SE^{(1)}} \rightleftharpoons \underset{b^{(2)}}{SE^{(2)}} \rightleftharpoons \ldots \rightleftharpoons PE$$

$$S+E^{(1)} \qquad S+E^{(2)} \tag{6.11}$$

Naturally, the resulting rate constant, k_b, is an essentially effective quantity.

If there is an excess of P in comparison with its equilibrium concentration, the direction of the overall process changes, the system passes through other non-equilibrium conformational states and can be described by (6.10), to which all the above reasoning is also applicable. Under the conditions of chemical equilibrium all steps become reversible and (6.9,10) can be represented by unified scheme

$$
\begin{array}{ccccc}
SE & \underset{k_{-b}}{\overset{k_b}{\rightleftharpoons}} & P\tilde{E} & \underset{k_{-c}}{\overset{k_c}{\rightleftharpoons}} & P + \tilde{E} \\
k_a \Updownarrow k_{-a} & & & & k_{-d} \Updownarrow k_d \\
S + E & & & & P + E \\
k_d' \Updownarrow k_{-d}' & & & & \Updownarrow \\
S + \hat{E} & \underset{k_c'}{\overset{k_{-c}'}{\rightleftharpoons}} & S\hat{E} & \underset{k_b'}{\overset{k_{-b}'}{\rightleftharpoons}} & PE
\end{array}
\tag{6.12}
$$

The concentration ratio of free S and P in equilibrium does not depend on the presence of an enzyme. But, the rate constant values of "mechanical" steps involving conformational changes (i.e., k_b, k_{-b}, k_d, k_{-d}, k_b', k_{-b}', k_d', k_{-d}') must diminish approaching the equilibrium. For a true mechanical system, equilibrium means rest. For such not strictly mechanical systems as protein macromolecules, one can expect a considerable retardation of these stages. Indeed it was shown [6.76] that, under conditions of dynamic redox equilibrium, one cannot register any conformation transition between the equilibrated forms of oxidized and reduced cytochrome C mole-

cules. We can assume that, for most enzyme processes, the relaxational mechanism
(and the general "machine-like" mechanism) of catalysis becomes ineffective, when
the system approaches the conditions of chemical equilibrium, and is replaced by
conventional chemical reactions between the catalytic groups of substrate and active center not involving compulsory conformational changes and the formation of
kinetically unequilibrated strained states directionally forcing the system along
the reaction path. It is, however, well known that most enzymatic processes within
the cell proceed under conditions which are far removed from chemical equilibrium,
and the advantages of "machine" mechanisms can be fully utilized.

6.6   Dynamic Model for Aspartate-Amino-Transferase

Many papers concerning the detailed mechanisms of certain enzymatic processes can
be found in contemporary literature. The more we know the more clear it becomes
that the functioning of enzyme constructions has a mechanical, machine-like nature.
In this section we shall discuss the results of this type of studies concerning
one of the best known enzymatic processes, the α-amino acids transamination dis-
covered by BRAUNSTEIN and KRITSMAN [6.77] in 1937

$$R_1-\overset{\overset{\text{H}}{|}}{\underset{\underset{\text{NH}_2}{|}}{C}}-COOH + R_2-\overset{\overset{}{}}{\underset{\underset{O}{\parallel}}{C}}-COOH \rightleftarrows R_1-\overset{\overset{}{}}{\underset{\underset{O}{\parallel}}{C}}-COOH + R_2-\overset{\overset{\text{H}}{|}}{\underset{\underset{\text{NH}_2}{|}}{C}}-COOH.$$

$$(6.13)$$

These reactions are catalyzed by aminotransferases with pyridoxal phosphate (PLP)

$$H_2O_3POH_2CH\overset{\overset{\text{H}\quad O}{\diagdown\diagup}}{\underset{\underset{\text{N}\quad CH_3}{}}{\overset{C}{\diagdown}}}OH$$

as a cofactor. Essential results clarifying the action mechanism of one of such
PLP-dependent enzyme, aspartate aminotransferase (AAT) were obtained in several
leading laboratories: those of Braunstein, Snell, Hammes and others. Our account
here is based on the excellent review of IVANOV and KARPEISKY [6.78]. These authors
have made an important contribution to the development of the model presented here.
Even if future investigations introduce certain corrections into some of the details of the proposed mechanism, one cannot doubt the authenticity of the general
picture.

In the model congruent reaction PLP serves as intermediary acceptor of the amino group, and, in the first reaction stages, is transformed through aldimine and ketimine into pyridoxamine phosphate (PMP)

$$R_1-\underset{\underset{NH_2}{|}}{\overset{\overset{H}{|}}{C}}-COOH \; + \; PLP \underset{+H_2O}{\overset{-H_2O}{\rightleftharpoons}} \text{aldimine} \longrightarrow \text{ketimine} \underset{-H_2O}{\overset{+H_2O}{\rightleftharpoons}}$$

$$R_1-\underset{\underset{O}{\|}}{C}-COOH \; + \; PMP \tag{6.14}$$

PMP then reacts with a keto acid $R_2 - \underset{\underset{O}{\|}}{C} - COOH$ and undergoes the same transfor-

mations in the reverse direction forming PLP and amino acid. AAT catalyzes reaction (6.13) between aspartic acid and α-ketoglutaric acid. The enzyme consists of two identical subunits each having the molecular weight of 45000. In the initial state (PLP) coenzyme forms an aldimine double bond with the ε-NH_2-group of one of AAT lysine residues (Fig.6.3). However, it is not only the carbonyl group of coenzyme that takes part in the bonding of the latter with protein. All the functional groups attached to the pyridine ring were shown to interact with one or another protein group. An important part seems to be played by the phosphate group. This group exists as a unicharged anion and interacts with one of the protein cationic groups. The nitrogen atom of the pyridine ring accepts a proton, probably from the OH-group of protein tyrosyl residue, and becomes positively

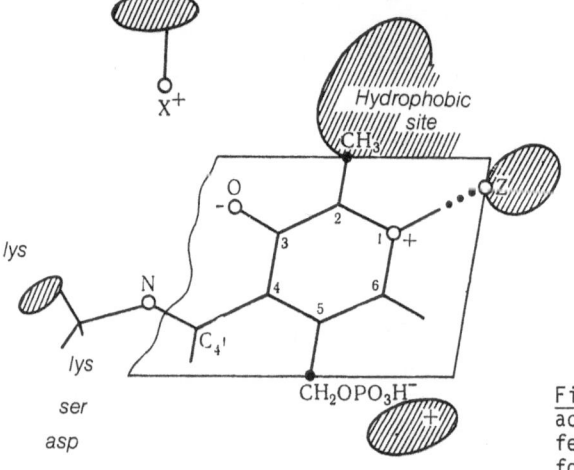

Fig.6.3. Pyridoxal phosphate in the active center of aspartate aminotransferase [6.79]. The shaded regions stand for protein-apoenzyme

charged (Sp^2-hybridization). The imine nitrogen, formed in such a way, interacts with the negatively charged oxygen atom (Z) of tyrosyl residue (a strong hydrogen bond).

There is convincing evidence that PLP methyl group interacts (Van der Waals forces) with the hydrophobic region of apoenzyme. The large decrease of the pK_a value of the PLP (in complex with AAT) oxygroup, in comparison with the pK_a value of this group in free PLP (pK_a 6.2 instead of 10.5), cannot be explained by the protonation of pyridine ring N-atom only. It means that an apoenzyme positively charged group (X^+) must be positioned in the neighborhood of the PLP oxygroup.

This group is, probably, a lysine residue (see Fig.6.3). In the equilibrium stable complex of PLP with enzyme, the interaction is, thus, realized by all PLP groups able to take part in binding ("multipoint binding", providing a unique fixation of PLP on the protein). If the initial reagents are free PLP and apoenzyme, the complexing then leads to a protein conformational change which allows all the above bonds to be formed and, simultaneously, modifies the coenzyme (aldimine covalent bond formation and proton transfer to the nitrogen atom of the pyridine ring).

Let us follow now the individual steps of enzymatic reaction (6.13) beginning from the free substrate-enzyme complex (1 in Fig.6.4).

The substrate α-amino acid (1-aspartic acid) forms an ionic complex 2 (the amino group of free amino acid is protonated, and the carboxylic group ionized in the physiological pH range). The negatively charged PLP oxygroup serves as a counterion to the substrate aminogroup, and one of the protein cationic centers, to its carbonyl group. The next step consists in nucleophilic addition of the substrate amino group to the internal double aldimine bond C=N that connects the PLP carbonyl group with the apoenzyme lysine residue. The formation of this bond requires the fulfillment of two conditions: a) the aminogroup must be deprotonated ($-NH_3^+ \rightarrow -NH_2$), and b) the distance between amino-group nitrogen and formil carbon of aldimine C=N group must be reduced from 3.5 A (the sum of C and N van der Waals radii) to 1.5 A. Condition (a) is fulfilled due to the change of pK values of the substrate NH_3^+-group and the PLP oxygroup by complex formation. IVANOV and KARPEISKY showed that the transfer of proton from the substrate NH_3^+-group to the PLP O^--group (3 in Fig.6.4) must be realized. Examination with the use of molecular models reveals that condition (b) is fulfilled through the rotation of coenzyme arround an axis connecting the 2-methyl and 5-methylenphosphate groups fixed by the apoenzyme (Fig.6.5). This rotation requires the rupture of PLP bonds with protein through the oxygroup and nitrogen atom of pyridine rings. The first bond is already ruptured due to the protonation of the $-O^-$-group [see condition (b)]. The protonation must also lead to the decrease of basicity of the ring nitrogen atom and, consequently, to the disruption of its hydrogen bond. The PLP bond with apoenzyme lyzine must, however, remain, but is transformed in the course of nucleo-

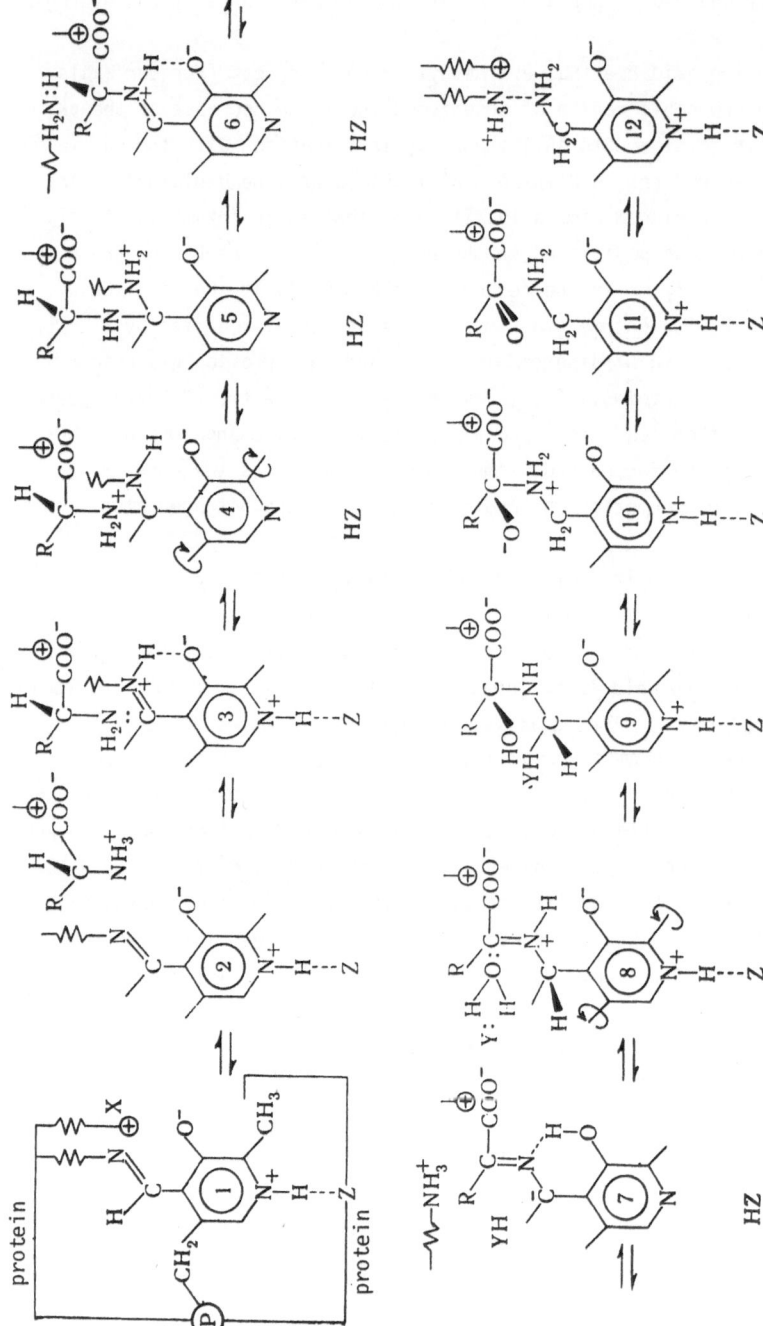

Fig.6.4. Scheme of chemical steps of enzymatic transamination (first part of reaction). According to [6.79]

philic substitution from a double into a single bond. Probably, it is just the re-laxational change in conformation, caused by the initial substrate-enzyme complex formation, that leads to the corresponding displacement of lyzine residue and, consequently, to the rotation of the PLP ring plane around the methyl-phosphate axis[8]. All the chemical changes, nucleophilic substrate attachment included, take place in the course of this conformational transformation (4 in Fig.6.4).

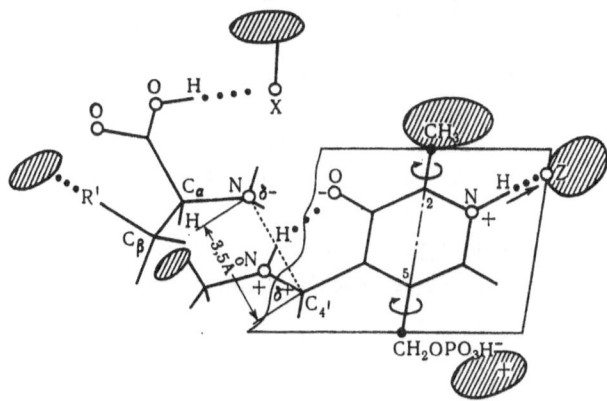

Fig.6.5. Structure of the enzyme-substrate aldimine complex 3 (see Fig.6.4), accord-ing to [6.79]. Arrows indicate the directions of rotation and displacements during step 4. Apoenzyme regions are shaded

A double aldimine bond must then be formed between the PLP carbonyl C-atom and the substrate amino group with a simultaneous disruption of the single bond between this C-atom and the lysine amino group (5 and 6 in Fig.6.4). It is, of course, pos-sible that, during the enzymatic process, the last two stages are realized in the course of protein conformational change without the fixation of states 5 and 6. Aldimine formation, thus, leads to the appearance of a free amino group in the apo-enzyme lyzyl residue in close vicinity of C-H bond. This facilitates the transfor-mation of aldimine group $>C = \overset{+}{N} < \overset{C}{\underset{H}{C}}\alpha$ into ketimine group $>\bar{C} - N = C_{\alpha}^{-}$ with proton transfer to lyzyl residue and carbanion formation (7 in Fig.6.4). It is, probably, the interaction with lyzine that leads to the conformational change which, in turn, results in the back rotation of the coenzyme ring plane around the methyl-phosphate axis (this is not discussed in [6.79] either). As a result, the carbanion is drawn together with YH—a proton-donating group of apoenzyme (Y is, probably, the imida-zole ring of a histidine residue), transferring its proton to the carbanion C^--atom (8,9 in Fig.6.4). The coenzyme fixation on protein by the rest of the bonds is simultaneously restored. The remaining steps (states 10-12) which lead in the end to keto acid separation and to PMP formation can be easily understood from Fig.6.4.

---

[8]This is not discussed in [6.78].

The second part of the process, beginning with PMP interaction with the second substrate-α-keto acid, requires that the PMP amino group be in neutral form and have a low pH value. This is provided for by the presence of strong cationic groups of apoenzyme, NH_3^+-group of lyzyl residue included, in the vicinity of the PMP amino group (12 in Fig.6.4). The steps of the second half of the process replicate the steps of its first half in reverse sequence.

The reader can find a more detailed description of structural changes accompanying each step, as well as experimental data verifying the ideas set forth, in [6.78]. It must be emphasized that the two principal steps of the whole process—the nucleophilic attachment of the substrate with covalent bond formation and the aldimine-ketimine tautomerization involving the carbanion formation stage—are realized due to certain macromolecular conformational changes providing for, with high precision, the necessary changes in the mutual orientations and separation of certain groups. It is just these mechanical steps that do not allow us to describe the enzymatic process on the basis of the classical chemical kinetics of gas reactions.

6.7 Conclusions

In this chapter new ideas concerning the physical mechanisms of elementary acts in enzymatic catalysis were presented. It is clear from the above that the problem of developing a quantitative physical theory of enzymatic catalysis is reduced to that of developing a quantitative kinetic theory for conformational relaxation of protein macromolecules and their complexes. Research in this direction has not yet been started.

It must be borne in mind that any quantitative theory (even if it is developed) cannot replace the existing and widely used (groundless, from my point of view) semi-quantitative empirical approaches based in the end, on the Arrhenius equation and the activated state theory. What researchers need is simple relations and parameters that would describe the kinetics of a process and have a clear physical meaning. Therefore, a task, more important than the creation of a quantitative physical theory of protein conformational relaxation, is to formulate new semi-empirical relations and to choose new parameters that, on the one hand, would assist in classifying and systematizing the experimental results and make it possible to plan kinetic experiments in a more meaningful way than is done at present, and, on the other hand, would somewhat more properly reflect the physical nature of the processes occurring in the course of enzymatic catalysis of biochemical processes.

To find such semi-empirical relations we, probably, need experimental studies of enzyme kinetics, mainly using the relaxation methods, able to characterize the reacting system in nonequilibrium conformations after an abrupt change of conditions (pH, temperature, substrate addition, etc.).

Investigations of this kind are now being carried out with many enzymes.

7. The Physics of Electron Transfer in Biological Systems

7.1 Overview

The problem of physical mechanisms of electron transfer in biological structures
and processes is one of the problems the solution of which, on the one hand, would
be of general importance for the understanding of the functioning of biological
systems, and on the other hand, can already be treated with the methods of biological physics. This problem is interesting not only in itself. It is closely connected with the general problem of energy transformation and accumulation in biological systems (see Chap.8).

Electron transport between redox carriers fixed in intracellular membranes has
for a long time been intensively studied in many laboratories throughout the
world. We shall discuss here the physical aspects of electron transfer between
fixed centers. Our discussion will be based on the experimental data concerning
the structure and functioning of electron transport chains in membranes of mitochondria and green plant chloroplasts and in chromatophores of photosynthesizing
bacteria. These electron transport chains play the most important role in such
basic processes as tissue respiration and photosynthesis. The discussion of the
hypothetical physical mechanisms of electron transfer in the above biological
systems will require detailed analysis of quantum-mechanical tunnelling conditions.

Acts of electron transfer, however, occur not only in electron transport chains.
Michaelis' works in the late twenties, and, especially, the development and widespread use of the electron spin resonance method in biological studies clearly
show that electron transfer accompanied by free-radical and ion-radical formation
is a compulsory stage of many biochemical reactions. Therefore, the rise and decay
of free-radical states, and the part played by these states in biochemical processes
will be discussed in a special section. Great attention will be given to the specific problems arising when one analyzes free-radical processes at the initial steps
of the light stage of photosynthesis (water oxidation and oxygen liberation), and
at the final steps of tissue respiration (utilization of molecular oxygen and water
formation). The mechanisms of these reversely directed processes requiring the accumulation of extremely active intermediates in the course of several consecutive
acts of electron transfer remain quite obscure up to now.

We shall deliberately omit here the multitude of works concerning the role of free-radical states and processes in radiation damages and other pathological states. I think that the existing vast experimental material is too intricate to become a subject of biophysical analysis.

During photosynthesis and, probably, during cell respiration, electron transport is preceded by processes of charge formation and separation. In the case of photosynthetic systems, these are the processes of pigment molecules neutral electronic excitations decaying to an electron and a hole localizing on the chemically active centers of the structure. For many years a discussion has continued in scientific literature about the part played in these processes by the semiconductor and photoconductor mechanisms. We shall devote a special section to this question and also to the more general problem of the applicability of concepts developed in solid-state physics to the description of biological systems.

It is convenient to begin with questions concerning the participation of free-radical and ion-radical states in biochemical reactions.

7.2 Free Radicals and Radical Ions in Biochemical Reactions

As early as 1929 MICHAELIS [7.1] studied redox reactions of certain metalloorganic complexes using the method of potentiometric titration and came to the conclusion that the processes of two-electron oxidation or reduction of ligands in these complexes proceed stepwise through two one-electron stages with the formation of intermediary semireduced (semioxidized) products. MICHAELIS suggested that these intermediate (he called them semiquinones) are of free-radical nature [7.2,3]. Similar intermediary steps were also found in redox transformation processes of many dyes and quinone compounds. MICHAELIS believed that one-electron steps are compulsory stages of all redox reactions, biochemical reactions included. Using spectral and magnetometric methods with model systems (dyes, quinones) MICHAELIS succeeded in finding additional evidence of the formation of intermediary semioxidized or semireduced products of the free radical type.

The first experimental work to reveal that semiquinones are formed during a biochemical process, was probably that of Haas [7.4], who found that, in the course of old yellow enzyme reduction, there appears an intermediary dyed compound (according to the correct suggestion of the author, a substrate-enzyme complex of the free-radical nature). After that, Theorell, Chance and other authors carried out important investigations of fast one-electron steps in enzymatic redox processes (see the excellent review by CHANCE [7.5]). From my point of view, however, the most important advance was made in the pioneer work of COMMONER et al. [7.6], where free-radical states in metabolizing tissues and cells of animal and plant origin were for the first time recorded with the help of the electron spin resonance method. These results were later verified by others [7.7-9].

Semiquinone formation during the stepwise oxidation or reduction can be most conveniently discussed with the mutual transformations of p-benzoquinone and p-benzohydroquinone taken as an example. These reactions can be written as follows:

Reduction

a Quinone ⇌ Semiquinone ⇌ Hydroquinone

$+e^-$ / $-e^-$ $+e^- +2H^+$ / $-e^- -2H^+$

Quinone   Semiquinone   Hydroquinone

Oxidation

b 2 [semiquinone] $\xrightarrow{+2H^+}$ / $\xleftarrow{-2H^+}$ [quinone] + [hydroquinone]

Disproportioning

c [semiquinone] $\xrightarrow{+H^+}$ / $\xleftarrow{-H^+}$ [semiquinone]

Acid-base transformations of semiquinone

d [hydroquinone] $\xrightarrow{-H^+}$ / $\xleftarrow{+H^+}$ [hydroquinone] $\xrightarrow{-H^+}$ / $\xleftarrow{+H^+}$ [hydroquinone]

Acid-base transformations of hydroquinone

(7.1)

Process (7.1a) is stepwise reduction of quinone (from left to right) and stepwise oxidation of hydroquinone (from right to left) through intermediary semiquinones in the form of radical ions. Reduction requires an electron donor, and oxidation an electron acceptor (usually molecular oxygen). The direct and back dispropotionation reactions (7.1b) are also to be taken into account in the general scheme of quinone and hydroquinone mutual transformations. Semiquinone and hydroquinone can exist in different ionization states, depending on pH [see (7.1c,d)].

Semiquinone is most stable in the form of radical ion, i.e., at high pH values. Therefore, working in the alkaline pH range, it is possible to reach radical-ion concentrations high enough to register intensive ESR signals. The kinetic and equilibrium constants of the processes of (7.1) type were measured by several authors (see, e.g. [7.10]). Investigations of such model redox systems using the ESR method have been described in detail in many monographs and reviews [7.11-14].

When the paramagnetic centers giving rise to an ESR signal of the organic radical and radical-ion types had been found in the tissues and cells of animal and plant origin, it was, first of all, naturally suggested that these centers are identical with semiquinones, formed in the course of stepwise oxidation or reduction of many biochemical compounds. Indeed, in cytoplasm and subcellular structures, there are many biochemically active substances (vitamins, coenzymes, and such electron carriers as naftoquinones, flavins, ubiquinone, etc.) whose oxidation and reduction must proceed in a stepwise manner with the formation of free-radical intermediates. The actual reality, however, has been found to be much more complicated. The mechanisms of formation and decay of paramagnetic particles giving rise to an ESR signal within the cells, their nature and function, are far from clear.

The majority ESR laboratories throughout the world have studied in the recent years the ESR spectra of intracellular paramagnetic complexes of metals, which play an important part in the processes of electron transport and energy transformation within membrane structures of mitochondria and other organelles. It is a pity that the number of works concerning the physical and chemical mechanisms of the formation and functioning of free-radical paramagnetic centers in the course of dark intracellular processes is much smaller, although these mechanisms may be of great biological importance. Let us discuss the principal data and formulate the questions as yet unanswered.

In the works carried out at the close of the fifties and the beginning of the sixties lyophylized (freezedried) tissues and cell suspensions were, as a rule, studied. The reason for this was the low sensitivity of ESR spectrometers. Freeze-drying of biological objects is usually carried out in the following way. The preparation is rapidly frozen in liquid nitrogen or solid carbon dioxide, and the water sublimated into the vacuum. It was postulated that in the dry powder (water content 1.5-2.0%) obtained by this method, the state of the preparation (concentrations of ingredients, macromolecular and supramolecular structures, etc.) is the same as that "trapped" at the moment of freezing. In freezedried samples, many authors observed ESR signals in the form of a slightly asymmetrical line with g-value close to 2.00, separation between the points of maximal slope $\Delta H_{max} \approx 8$ Gauss, and integral intensity corresponding to $\sim 10^{17}$ spins/gm. With the progress in ESR techniques began the studies of water containing humid tissue samples [7.15-17]. It was found that native preparations without liophylization also gave free-radical ESR signals, but their integral intensity in analogous objects is always less by about one order of magnitude, and the width is approximately twice as large ($\Delta H_{max} \approx 14-16$ Gauss).

The changes in the concentration of free-radical states and in the shape of their ESR signals cannot be attributed to the freezing process: it is well known that ESR signal parameters of frozen tissues do not differ from those of free-radical centers in native preparations [7.18,19]. It, thus, seems that the changes of these parameters occur as a result of the freezedrying process. For many years in our laboratory we have carried out studies aimed at clarifying the origin and nature of these changes. I shall discuss these studies in greater detail. They are interesting not only in themselves, but in connection with the questions which arise from the obtained results, and which have had no final answers up to now.

The principal data concerning the freezedrying process have been listed in [7.20,21]. From the beginning, it was shown that in freezedried samples one can obtain free-radical ESR signals identical to those in native or frozen samples. Firstly, it is necessary to follow strictly the proper freezedrying techniques (see [7.22]). The main requirement is the speed of the freezing and the maintenance of low temperature until the water content diminishes to the level of 4-5%. Secondly, even a momentary contact of the lyophilized sample with atmospheric moisture must be completely excluded (this is essential to isolate the sample from the water and not from the oxygen). In these conditions, freezedried preparations exhibit a free radical ESR signal ($g = 2.0049$, $\Delta H_{max} \approx 14.5$ Gauss) undistinguishable from ESR signals of native or frozen tissue preparations (Fig.7.1, curves A and B). Even a momentary contact with humid air leads to the appearance of another ESR signal ($g = 2.0057$, $\Delta H_{max} \approx 8$ Gauss) superimposed on the "native" signal. The intensity of the signal whose $g = 2.0057$ increases with the duration of contact with moisture (Fig.7.2). After that, these paramagnetic centers decay slowly and, in the end, irreversibly. Thus, the free-radical paramagnetic centers responsible, in animal tissues, for ESR signals with $\Delta H_{max} \approx 8$ Gauss and $g = 2.0057$ rise as a result of the contact of lyophilized preparation with humid air. Freezedrying, and to be more precise, the decrease of the sample water content to a level lower than 15-20% is an obligatory requirement for these centers to be formed. The same can be observed in the case of plant tissue preparations [7.23]. In a prelimenary way we can, thus, conclude that dehydration in the course of freezedrying leads to changes in cellular structures, due to which considerably greater quantities of quinone compounds undergo the stepwise oxidation and reduction with free radical formation as compared with those in cells undamaged by the freezedrying process. Apparently, paramagnetic centers responsible for the "native" signal with $g = 2.0049$ rise, mainly, in mitochondria [7.24] and paramagnetic centers responsible for the "artefact" narrow signal of freezedried tissues with $g = 2.0057$ are, mainly, localized in the cell plasma [7.25].

Analysis of the data concerning the "artefact" free-radical centers in lyophilized preparations made it necessary to formulate the unforeseen and important biophysical problems which have not been solved up to now.

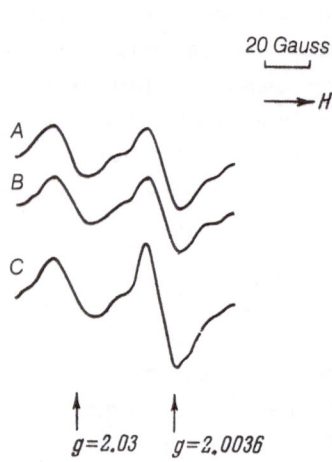

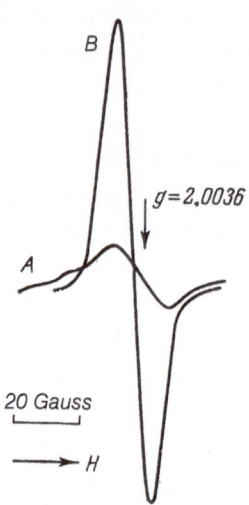

Fig. 7.1. ESR spectra of rat liver preparation [7.21]. (A) frozen preparation;
(B) "properly" lyophilized preparation;
(C) sample "B" after momentary contact
with humid air. Signal g = 2.03 is that
of non-heme-iron protein complex in mitochondria

Fig. 7.2. ESR signal of "properly"
lyophilized rat liver preparation before
(A) and after (B) 24-hour contact with
humid air [21]

It is possible to model the principal characteristics of ESR signals of lyophilized tissues with the help of various radical ions adsorbed in proteins
[7.26-29]. Proteins selectively adsorb radical ions (semiquinones) shifting the
equilibrium (7.1) towards their formation, and the ESR signals of adsorbed radical
ions resemble those of lyophilized tissues. Important results were obtained using
the so-called "vapor-flow" method [7.30]. This method allows one to keep the sample
(a tissue or model compound adsorbed in a protein) in a given atmosphere with a
fixed water content within the sample and in the environment. For a long time semiquinones in tissues, as well as in model systems, were believed to rise and decay
only due to process (7.1a) going on from the right to the left, i.e., due to stepwise oxidation of reduced carriers of the hydroquinone type with molecular oxygen
as electron acceptor (oxidizer). Certain facts obtained in the works cited above
[7.20,21] do not, however, agree with this assertion. Investigation of lyophilized
tissue preparations using vapor-flow techniques showed that, if organic solvent
vapors are added to the water-free[1] carrier (O_2), the presence of oxygen does not
lead to the rise of artefact ESR signal, and, in general, does not affect the
sample properties (Fig.7.3). At the same time, the use of water vapors with an inert
carrier (argon) leads to the rise and subsequent irreversible decay of "artefact"
ESR signals, the decay in argon atmosphere being considerably faster than in O_2

[1]Special precaution is necessary when operating with extremely dry gases and vapors.

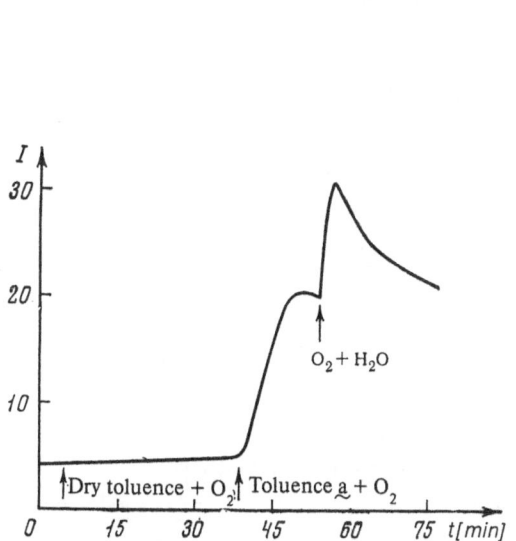

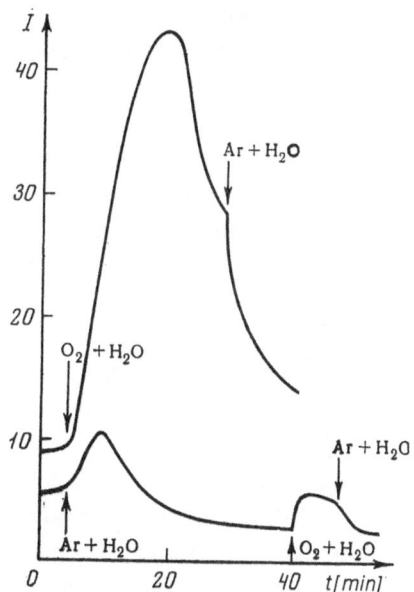

Fig. 7.3. Effect of water admixture in toluene on the rate of artefact ESR signal rise. Study of rat liver preparations by vapor-flow method [7.20]

Fig. 7.4. Irreversible decay of "artefact" free radicals in lyophilized rat liver preparation due to water vapors in the inert carrier [7.20]

atmosphere. This can be seen in Fig.7.4. An insignificant signal rise (low curve) after the replacement of Ar by O_2 is due to the rise in the "native" mitochondria signal intensity ($\Delta H_{max} \approx 14$ Gauss). Certain kinetic characteristics of the rise and decay processes of "artefact" paramagnetic centers in lyophilized tissues can be reproduced using a model system: alkaline filter paper with adsorbed benzoquinone. As in the case of lyophilized tissues, O_2 slows down the decay of semiquinones, and, therefore, if the experiment is conducted in O_2, the intensity maximum of the ESR signal is higher than it is in Ar atmosphere. It was also shown that the formation of "artefact" paramagnetic centers in lyophilized tissues and their gradual irreversible decay take place only if the sample water content is reduced in the course of lyophilization to a level lower than 20%, i.e., if it does not exceed the quantity of water that is rigidly bounded by cell protein structures. The analysis of experimental data leads to the following conclusions (see, e.g. [7.19]).

1) The "artefact" signal of lyophilized tissues is produced by semiquinone radical ions, arising not only due to one-electron oxidation of completely reduced electron donors, but also due to one-electron reduction of completely oxidized species of quinone compounds within the cell plasma.

2) One-electron reduction is accompanied, in this case, by a gradual irreversible decay of the ESR signal.

3) The formation and irreversible decay of radical ions become possible as a result of some destructive processes due to the decrease of cell water content to 12-15%.

4) One of the main sources of radical ions of this type is ascorbic acid [7.31]. The parameters of the ascorbic acid radical-anion ESR signal coincide with those of the "artefact" ESR signal in lyophilized tissues.

Studies carried out recently in our laboratory have shown kinetic properties of the "artefact" ESR signal to depend on redox potential changes within the cells, which, in turn, are caused by the changes of mitochondria oxidative activity and oxygen diffusion velocity with water content.

The conclusions formulated above allow us to ask a question: what serves as electron donor in the processes of reduction of completely oxidized compounds in lyophilized tissues and model systems? In connection with this question, it is necessary to consider the results obtained in the studies concerning electron transfer reactions between water and organic electron acceptors in model systems.

Interactions of nucleofilic agents (hydroxyl ion, alkoxianions, etc.) with organic electron acceptors (quinones, dyes, nitrocompounds, tetracyanethylene, tetracyanquinodimenthan, etc.) are known to lead to radical-anion formation [7.10,32-39].

In the case of the strongest acceptors (e.g., tetracyanethylene), the formation of radical anions, A^-, can be observed when electron acceptor, A, interacts with water even at neutral pH values. In most cases, however, it is necessary to use alkaline solutions. The reaction rate and the attainable concentrations of radical anions A^- (which decay more or less rapidly, due to secondary processes) increase with OH^- concentration. This fact, as well as the identity of reaction characteristics for quite different electron acceptors, have led to a suggestion that the primary act of electron transfer is realized as follows [7.33]:

$$OH^- + A \rightarrow OH + A^- \ . \tag{7.2}$$

Reaction kinetics did not contradict this scheme [7.36,40,41]. It was, however, always possible to assume the existence of a chemical reaction (specific for every type of A), in the course of which radical anions arise due to secondary processes after initial OH^- nucleophilic addition to A [7.10,42]. In order to prove that the primary elementary act is one-electron transfer (7.2), it is necessary to reveal the OH radical.

In the case of weak electron acceptors, the formation of radical anions cannot be observed even at very high pH values. This process, however, goes easily under the action of illumination with wavelengths lying within the interval of the acceptor absorption spectrum. The electron was shown to be accepted by A molecules in the electronically excited triplet state [7.36]. For these states, electron transfer from hydroxyl ion to A is thermodynamically favorable, and, if stage (7.2) is realized, the OH radical must appear as a kinetically independent particle.

OH radicals in these systems were indeed recorded by chemical methods [7.43]. The kinetics of their formation coincides with that of radical anions A⁻ and, with the help of the method used, it was possible to get the OH radical yield of up to 80% of the theoretical value corresponding to (7.2).

Thus, in the case of a photoinduced process, the act of one-electron transfer from the OH⁻ ion to the organic electron acceptor molecule really takes place.

This simple scheme does not, however, hold true for dark processes. Electron transfer from the OH⁻ ion to acceptor is, as a rule, thermodynamically unfavorable, and the process (even taking into account subsequent favorable reactions of the OH radical) should go on so slowly that it could not be recorded. A scheme was proposed and experimentally verified, according to which in dark processes (with the exception of extremely strong acceptors, such as tetracyanoquinodimethane) the first step includes the simultaneously proceeding acts of electron transfer from OH⁻ to A and of fixation of the formed OH radical by a third particle. This third particle may be another molecule of the acceptor, or a molecule of the radical anion, or of any organic compound [7.44,45]. OH-radical fixation is realized either as a result of addition to a double bond, or by way of dehydrogenation. In both cases, there arise active free radicals undergoing subsequent transformations which, in the end, lead to the irreversible decay of semiquinones and their precursors. These mechanisms are, probably, quite common in organic chemistry, particularly in the cases of nucleophilic and electrophilic substitution reactions. Research in this field is carried out in many laboratories [7.38,46-48].

The assumption, according to which the water (or OH⁻) serves, in lyophilized tissues, as one of the electron donors to the molecules of quinone type, appeares absurd at first sight. In model systems, the measureable rate of electron transfer between OH⁻ and A can be observed in the dark only at high pH values (pH > 9-10). This requirement is always fulfilled when benzo- and naphtoquinones, flavines, and other intracellular electron acceptors are used as models. It was, experimentally shown, however, that, in the presence of proteins, free radicals of the semiquinone type can be formed by one-electron transfer between OH⁻ and quinone molecules even at physiological pH values [7.49]. This may be caused by the local increase of effective pH value at the protein surface, by the stabilization of radicals on protein, and by electron transfer according to Grottgus mechanism between separated particles through water bounded by protein.

Addition of quinones to lyophilized tissue preparations leads to the formation of radical ions of these quinones under the action of water. The patterns of this process are very similar to those of the rise and decay of intrinsic "artefact" free-radical centers in lyophilized preparations.

These data on the one-electron transfer between water (OH⁻) and organic electron acceptors in model systems and "improperly" lyophilized tissue preparations force us to pose a question about the existence of analogous processes in native

biological structures. In other words, can water (OH⁻ ions) take part in a biochemical redox reaction as an electron donor? (Naturally, we are not speaking here
of photoinduced oxidation of water in photosystem II of higher plants and green
algae). This assumption was made in [7.50] on the basis of results obtained in the
study of the correlation between the free-radical ESR signal in mitochondria and
oxidative phosphorylation. Certain experimental data led to a hypothesis (absolutely
unproved) that, in the mitochondria respiratory chain, half of the electrons going
through the cytochrome system to molecular oxygen are delivered not by the oxidized
substrate but by water. Let us temporarily assume that this hypothesis is true.
What must be the fate of OH radicals which are to be formed? The only way to get
rid of these radicals, without detrimentally affecting the cell, is to form molecular oxygen. It meas that these radicals must, at least in pairs, get into the
"chambers" preventing their contact with organic structures within the cell, and,
after that, either recombine with hydrogen peroxide being formed or undergo further
oxidation to molecular oxygen, each of them delivering one electron to the chain
(and one proton to the medium). In the first case, hydrogen peroxide must be de-
composed by catalase[2]. In any case, transfer of four electrons from water leads to
the liberation of one oxygen molecule, i.e., compensates exactly the odd O_2 absorbed
due to these electrons being transferred to cytochromeoxidase.

It is, thus, impossible either to prove or to disprove this improbable suggestion
when studying the summary O_2 absorption.

In living nature, however, there exist systems in which water molecules undoubtedly play the part of electron donors. We are speaking of the processes of photoinduced water oxidation with molecular oxygen liberation in leaves of green plants
and in unicellular algae. This reaction proceeds (see, e.g. [7.51]) in photosystem
II, the electron acceptor being, probably, the radical ion of chlorophyll, $Chla_2^+$,
forming, as a result of the photoionization of photosystem II active center. The
electron is transferred from the water to this center not directly but via a system
of carriers including, probably, Mn^{2+} [7.52] and Cl^- [7.53] ions. As a matter of
fact, there is no mystery in the photooxidation of water during photosynthesis.
Practically, in any model system with dyes and other organic electron acceptors,
the excitation of the latter by light leads to the photooxidation of water (see
above). Electron affinity of the excited acceptor molecule, or that of the radical
cation formed by photoionization, is almost always high enough to make electron
transfer from the water molecule (or from OH⁻) thermodynamically favorable. In
model systems, however, the photooxidation of water is not as a rule accompanied
by oxygen liberation. The chemical activity of OH radicals (or their ionic forms
H_2O^+ or O^-), produced as a result of electron transfer, is so high that they im-
mediately react with other chemical compounds (first of all, with the organic

---

[2]It would explain the general presence of catalase within the cells.

electron acceptor itself) and induce various, and uncontrollable, chemical transformations. Thus, the mystery lies not in the photooxidation of water, but in the
mechanism of OH-radical transformation into molecular oxygen. The question is: in
what way are the OH radicals, formed in the course of water photooxidation in a
system containing an immense amount of various organic compounds, transformed into
O_2 with practically, a 100% yield? [7.54]. How does a plant cell defend itself
against OH radicals? Photosynthetic bacteria—evolutionary the most ancient photosynthetic species—do not have this problem. As a result of photooxidation their
primary electron donors form much less active free radicals (e.g., SH radicals) that
can be transformed into end products (e.g., into free sulphur) which do not damage
intracellular compounds. These species possess only one photosystem, and the transfer of one electron from the primary donor to the final acceptor requires the absorption of one quantum only. When, in the course of biological evolution, photosynthetic organisms turned from "expensive" exotic donors into "cheap" widely distributed water "fuel", it was accompanied by essential rearrangement of the photosynthetic apparatus. There appeared photosystem II, specially designed for water
photodecomposition, and the transfer of one electron came to require two light
quanta with a corresponding efficiency lowering. It can be assumed that transition
to the two-quantum mechanism, conditioned by the development of photosystem II, was
caused not by the fact that water photooxidation requires more energy than could
be delivered by the quantum absorbed by the active center of photosystem I, but,
mainly, by the necessity of spatial separation of water photooxidation sites from
the chemically vulnerable cellular components. The difficulty of the problem is
redoubled by the fact that O_2 formation requires at least two OH radicals, rising
in close vicinity of each other, but not simultaneously. (An oxygen molecule may
be formed either from four water molecules, each delivering one electron to the
acceptor, or from two water molecules, each consecutively delivering two electrons).
It seems that the construction within photosystem II, providing for OH-radical
stabilization, also includes Mn^{2+} ions, whose presence is obligatory for O_2 liberation.

The "invention" of the OH-radical stabilization method, the "elaboration" of
a suitable molecular construction, was one of the most important events in the
course of biological evolution in our planet. Apart from leading to molecular oxygen
formation in the atmosphere, this "invention" has made it possible for aerobic
organisms using the systems of tissue respiration to appear. The terminal stages
of tissue respiration, during which O_2 is reduced to water,—a process reverse to
photosynthetic water oxidation—must also pass through the steps of OH-radical
formation. The first step in O_2 reduction is the formation of radical ion O_2^- (the
anion of radical HO_2). It seems that, in the cases when molecular oxygen is reduced
but the universal construction for the stabilization of OH radicals and their transformation into water is absent, the role of this construction is played by special
enzymes—HO_2-dismutases—widely distributed among anaerobic and aerobic organisms

[7.55,56]. It has been suggested that active centers of these enzymes include Mn^{2+} ions (E. Coli [7.57]) or Cu^{2+} and Zn^{2+} ions (animal tissues [7.58]).

A unique laboratory process in which water oxidation leads to stoichiometric liberation of O_2 is, probably, electrolysis with a nonoxidizable anode. In this case, the formed OH radicals "wait for each other", and, are in the end, transformed into molecular oxygen. An analogy suggests itself, and ideas of an electrolytic mechanism of water decomposition in photosynthesis were presented [7.59-61]. In model photocells [7.62], dyes — chlorophyll analgoues — were used giving potential differences high enough to provide for water electrolysis. I think, however, that these ideas and works, very important with respect to practice, are not directly connected with the true problem of photosynthetic oxygen liberation. One can certainly build a photobattery (using organic systems or selenium photocells, it does not matter which) and close the circuit through electrodes of an electrolyzer. The essence of the problem is not the origin of the energy necessary for water decomposition, but the design of a construction that stabilizes and isolates OH radicals in a chemical system of active compounds. A cell has no macroscopic spatially isolated electrodes made of noble metals. One can, of course, give the name of "electrode" to the hypothetical construction of several metal ions, where, presumably, OH radicals are trapped, but this cannot clarify the problem.

KRASNOVSKY and BRIN [7.63] observed oxygen liberation when they illuminated the electron acceptor (ferricyanide) adsorbed on ZnO surface in water suspension. These experiments were reproduced with organic electron acceptors (quinones), and it was shown (using H_2O^{18}) that oxygen is liberated in the course of water photodecomposition [7.64]. Thus, in this model system (having no resemblance to photosynthetic structures), OH radicals are stabilized and transformed into O_2 in the presence of organic compounds, which should, in principle, catch them. Perhaps, investigation of such models will help in solving the puzzle of molecular oxygen photosynthetic formation from water. The material presented in this section also shows that one-electron transfer acts must serve as intermediary steps in many biochemical reactions which have not been usually considered as free-radical reactions. It must, however, be borne in mind that the instantaneous concentrations of free-radical and ion-radical states may be too low to be recorded. About 14 years ago it was suggested [7.65] that, during oxidative phosphorylation, an intermediary semireduced radical-ion form of adenine ring appears, although ADP phosphorylation is generally considered as being a reaction of the acid-base type. BRZHEVSKAYA et al. [7.66-68] in a series of very interesting works have shown that, in the course of back enzymatic reaction — ATP hydrolysis by myosin — intermediary free-radical states arise, caused, probably, by the one-electron transfer steps.

7.3 Electron Transport Chains (ETC) in Mitochondrial and Chloroplast Membranes

Many aspects of the processes of electron transfer between redox centers fixed in mitochondrial and chloroplast membranes cannot be discussed without touching upon the problem of liberated energy conservation, i.e., the problem of oxidative phosphorylation. These questions are considered in Chap.8. Here we shall discuss electron transfer problems which can be analyzed independently of energy conservation.

In spite (and, may be, in consequence) of the already great, and becoming greater with every year, number of publications devoted to electron transport within subcellular structures, there are rather few strictly established facts and generally accepted ideas. We begin our discussion with electron transport chains (ETC) in mitochondria. All the principal ETC characteristics seem to be the same in all cells and organisms. By "ETC" we mean the sequence of carriers and processes realizing electron transfer from NADH or succinate to molecular oxygen. Coenzyme NAD (nicotinamide-adeninedinucleotide) is a mobile carrier which can be enzymatically reduced by various compounds, components of Krebs cycle included, with the exception of succinic acid, which is able to deliver electrons into the respiratory chain without NAD. The total reaction of NAD reduction may be written as follows:

$$NAD^+ + e^- + H^+ \rightarrow NADH \ . \tag{7.3}$$

Isotopic experiments have shown that, in the case of NAD^+ reduction by ethanol with alcohol dehydrogenase, the hydrogen atom passes directly from ethanol molecule to NAD^+ molecule. It is probable that NAD^+ interacts with other substrates in the same way. In the course of subsequent steps of electron transport along the ETC, protons do not pass directly, but into (or come out of) the water medium, whenever necessary.

Reaction (7.3) probably proceeds via two one-electron stages, but the steady-state concentration of intermediary free NAD (or $NADH^+$) radicals is so low that up to now they could not be detected.

We do not even know as yet the composition of mitochondrial ETC. Different scientific schools hold different opinions on the participation of certain components of the mitochondrial inner membrane in ETC functioning, and on the sequence of carriers in certain parts of ETC. We shall not discuss here the immense experimental evidence (mainly of biochemical nature) in favor of various points of view. These differences are not essential for the formulation of questions of more general biophysical importance. Therefore, we have chosen as the basis for our considerations the scheme proposed by CHANCE in one of his reviews [7.69] (Fig.7.5).

The carriers NADH, flavoproteins (FP) and ubiquinone (UQ) can, in principle, accept and deliver two electrons: their totally oxidized and totally reduced forms differ by two electron equivalents. Other ETC components are essentially one-electron carriers. Perhaps, the carriers of the first type, especially FP and UQ, function in ETC as one-electron carriers, passing totally oxidized and totally reduced

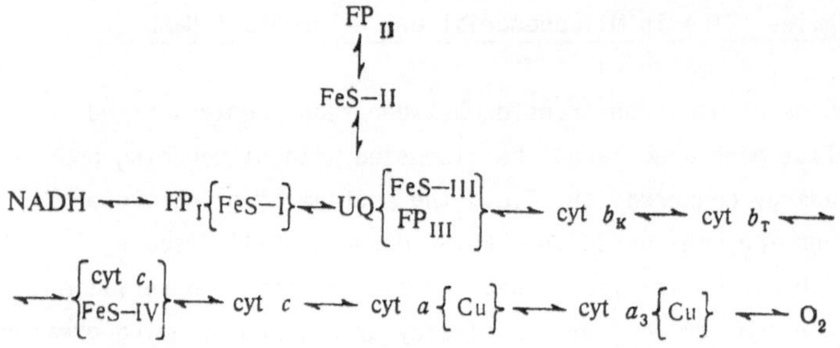

138

$$FP_{II}$$
$$\uparrow$$
$$FeS-II$$
$$\uparrow$$

$$NADH \longleftrightarrow FP_I\{FeS-I\} \longleftrightarrow UQ \left\{ \begin{matrix} FeS-III \\ FP_{III} \end{matrix} \right\} \longleftrightarrow cyt\ b_K \longleftrightarrow cyt\ b_T \longleftrightarrow$$

$$\longleftrightarrow \left\{ \begin{matrix} cyt\ c_1 \\ FeS-IV \end{matrix} \right\} \longleftrightarrow cyt\ c \longleftrightarrow cyt\ a\{Cu\} \longleftrightarrow cyt\ a_3\{Cu\} \longleftrightarrow O_2$$

Fig. 7.5.
Scheme of
mitochondrial ETC

forms through an intermediary semiquinone state (see Sect.7.2), or only between semiquinone and one of the end forms. Thus, flavoprotein, which transfers electrons from NADPH to cytochrome C in microsomal ETC, undergoes redox transitions between semiquinone and totally reduced forms of its active center [7.70]. Let us consider the carriers of Fig.7.5 in consecutive order. The active center of FP is flavine-ademinendinucleotide (FAD) — an isoalloxazine derivative. Below is shown the FAD transition from the quinone into the hydroquinone form:

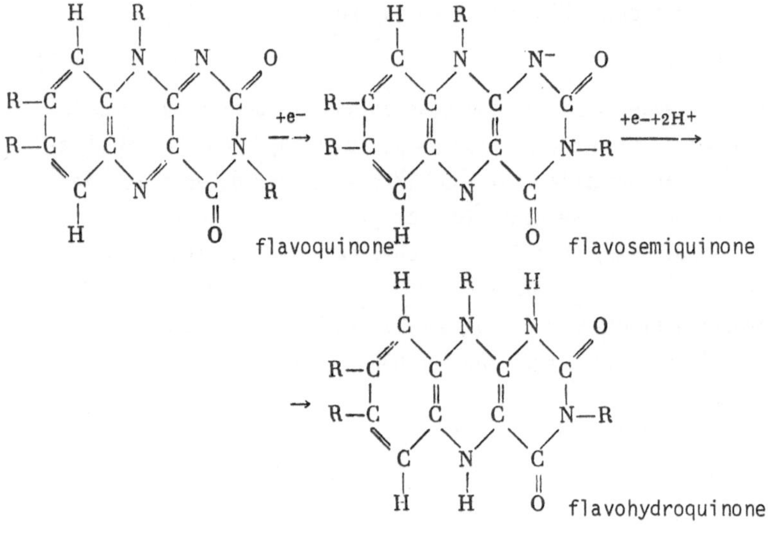

flavoquinone, flavosemiquinone, flavohydroquinone

Flavosemiquinones, arising during stepwise redox transformations of flavines, were shown to interact strongly with transition group metals tightly bound to proteins (see, e.g. [7.71-73]).

Many flavoenzymes contain transition group metals. Flavoproteins in mitochondrial ETC are, evidently, connected with ferroproteins. It is not without reason that multienzyme complexes, arising during ultrasonic fragmentation of mitochondrial membranes, always contain flavoenzymes and ferroproteines (these so-called "Green complexes" will be described later). The catalytic properties of purified flavoproteins differ slightly from those of undamaged mitochondrial ETC. Small changes in the extraction procedure (e.g., temperature increase from 30 to 37 C) change the enzyme's ability to interact with certain acceptors. The action of certain inhibitors on flavoproteins is influenced by their extraction from ETC (see, e.g. [7.74]).

Protein complexes with non-heme iron FeS (I,II,III,IV) are obligatory components of ETC. We do not know very much about them (more precisely, about their role in electron transport). This can be explained by technical difficulties of the registration of their redox changes in undamaged membrane structures. Essential progress made in this field in recent years is connected with the ESR method. We know now certain features of the structure of these carriers, it turned out that iron atoms are bound to sulphur groups of specific proteins. Indirect evidence has been presented concerning the participation of non-heme iron proteins in electron transfer and energy conservation processes within the ETC [7.75-79]. We shall later see that an analogous protein, ferredoxin, indeed, plays an important role in chloroplast ETC. Mitochondrial membranes contain a greater number of non-heme iron centers than those in cytochromes. Nevertheless, up to now one can find in the literature more experimental and theoretical papers on cytochrome enzymes than on non-heme iron complexes. In various particular theories the functions of energy transduction are ascribed to concrete cytochromes [7.69]. I think that the main reason of this is the habit of conducting scientific research in the field where there is more light and so it is easier to look. Sharp cytochrome spectra in visible and UV regions, where the pigment redox changes are easily detected, facilitate research to a considerable extent. There is no evidence, that non-heme iron protein complexes play a less important part than cytochromes in electron transport. Although all the known ETC components function as one-electron carriers, it does not exclude the possibility that the transfer of two electrons along parallel carrier chains of different composition is necessary for proper ETC functioning [7.80]. It should be borne in mind (see later), that, perhaps, for one ATP molecule to be synthesized, two electrons must pass through the energy transforming carrier of ETC. Iron-suphur proteins are tightly bound to various ETC regions and may, at certain sites, play the role of these parallel carriers. Convincing evidence of an iron-sulphur protein being necessary for energy transduction at the site between NADH and Q can be found in [7.81][3].

Flavoprotein FP_{II} (succinate dehydrogenase) realizes (probably, with iron-suphur protein FeS-II) the connection between the redox pair succinate-fumarate and the main ETC. All that has been said about FP_I can also be applied to this flavoprotein (and, of course, to flavoprotein FP_{III} with intensive fluorescence), which seems to form a redox pair with cytochrome b_k, but is not a component of ETC.

---

[3]New important results concerning the structure and function of iron-sulphur proteins have been recently obtained with the help of ESR study of mitochondria, tissues and isolated metalproteins at helium temperatures. New proteins taking part in electron transfer and energy transduction processes were identified [7.82, 83,200].

Ubiquinone (UQ) or coenzyme Q serves, according to certain assumptions (see, e.g. [7.84]), as, a mobile carrier not fixed in the membrane structure. These assumptions are based on the fact that UQ can be extracted from mitochondria in a low-molecular form not bound by proteins and lipoproteins. This last fact cannot, naturally, be regarded as any conclusive proof. We have already seen that UQ must be isolated from water in hydrophobic parts of the membrane. There is no reason to consider UQ to be more, or less, mobile within the membrane than other redox carriers. We shall later see that cytochrome C is as fixed in the membrane structure as other cytochromes, although its extraction proceeds even more easily than that of UQ. Incidentally, one cannot be sure of the functional and structural homogeneity of mitochondrial ubiquinone. In the "decoupled" mitochondria, having maximal electron transport velocity, the rate of NADH oxidation is somewhat higher than that of UQ oxidation [7.85]. It is possible that only a fraction of coenzyme UQ, whose total molar content in mitochondria is considerably greater than that of other carriers, participates in ETC electron transport [7.86].

Many properties of mitochondrial UQ change with the transition from undamaged organelles to the products of their more or less mild decomposition, to various submitochondrial particles extensively studied in the recent years.

Cytochromes (b_k, b_T, c_1, c, a, a_3) are chromoproteins whose prostetic groups are hemes, i.e., ferroporphyrins distinguished by substituents in the porphyrin ring. Cytochromes b_k and b_T contain chemically identical protoheme complexes of iron with protoporphyrin IX. The division into b_k and b_T is caused by small spectral differences, by differences in redox transformation kinetics and in the effect of uncouplers—antimicin A (an inhibitor), ATP and ADP—on the ratio between reduced and oxidized forms and on the kinetics of redox transition [7.69,80,87]. The data indicating the dependence of the state and functioning of cytochrome b_T on the state of the neighboring carrier (cytochrome c_1), are very interesting.

The prostetic group of cytochrome c_1 is the same as that of cytochrome C (see Chap.5), but protein molecular weight per one prostetic group is 51000, i.e., considerably greater than for cytochrome C. It is not known how many such units contain the cytochrome C_1 "ensemble" within mitochondria. This cytochrome is extracted in the polymeric form with molecular weight of 360000, i.e., as a polymer, containing about 7 subunits. However, since the extraction was carried out with a detergent, this last fact is not of great significance.

Much more is known about cytochrome C than about other carriers (see Chap.5). The easiness of its extraction with hypotonic salt solutions cannot be regarded as evidence of its being a mobile carrier, not fixed (in contrast to other cytochromes in ETC) in the membrane structure.

For a long time it has been known that the spectrum of native ferricytochrome C in solution is characterized by an absorption band at 6950 Å [7.88]. This band disappears as a result of heat treatment at alkaline pH values, and its disappearance is accompanied by cytochrome C losing its ability to be reduced by ascorbic acid, hydroquinone and some other reducing agents [7.89-91].

The appearance of band 6950 Å and the reducibility with certain reducing agents seem to be connected with the preservation of a specific native conformation (or configuration) in ferricytochrome C [7.92,93]. CHANCE et al. [7.94] have shown the mitochondrial cytochrome C spectrum to have the 6950 Å band, but this does not disappear with heat treatment sufficient for this band disappearance and cytochrome inactivation in solution. This result indicates that the interaction between cytochrome C and the surrounding membrane structure stabilize the native conformational state of cytochrome, influencing the ability of the macromolecule to undergo conformational transitions. We shall see later that the same conclusion can be drawn from the kinetic data.

Cytochromes a, a_3 and copper ions form an enzyme complex called cytochromeoxidase (for a long time it was considered to be a single protein). Isolation and careful purification of cytochromeoxidase lead to an enzymatically active (i.e., able to oxidize cytochrome C) pentamer, containing five protein subunits of 72000 m.w. each. A subunit contains one heme, "a", one copper atom and one atom of non-heme iron [7.95,96], but is enzymatically inactive. In ETC, as well as in isolated cytochromeoxidase, the a and the a_3 cytochromes differ kinetically and spectroscopically. It is, probably, caused by the difference in their packing and, consequently, in their conformation. The ratio of heme a and a_3 molar concentrations in ETC equals unity. Isolated cytochromeoxidase is not totally equivalent to the corresponding part of ETC: kinetic relations between cytochromes a and a_3 are different [7.97]. Copper ions are tightly bound to heme proteins, and undergo redox transformations during the functioning of ETC [7.98,99].

Before turning to the analysis of thermodynamic and kinetic data on ETC functioning, it is necessary to discuss two questions of general nature. The first question can be formulated as follows: is it possible to single out concrete fixed ETC in the inner mitochondrial membrane? In other words, does electron transfer within the ETC proceed between specific molecules, or the transport chains may be branched at various sites and, therefore, we must speak not of the electron transport chain but of the electron transport net? In the latter case, ETC stoichiometric composition is definite only on the average. As a matter of fact, this formulation of the question, as any drastic statement in science ("either, or") does not make it possible to give a definite answer: the answer depends on the concrete objective of the investigation, on the characteristic time intervals of the processes studied. This problem was carefully considered during the special and very stimulating discussion at the Symposium on Oxidases and Related Redox Systems [7.100]. CHANCE produced convincing evidence of chain branching taking place between cytochrome C and O_2 sites in the ETC but with rates by 2-3 orders of magnitude lower than that of electron transfer along the chains. A possibility is not excluded that, at the UQ

site, the branching is faster, but, probably even in this case, every chain has
its own structurally determined ubiquinone pool attending to the given chain much
more effectively than to the others. In any case, considering the kinetics of electron transfer along the cytochrome chain to O_2 (and, probably, along the whole chain
from NADH, or succinate, to O_2), we are justified in assuming the transfer to be
taking place between fixed carriers and in neglecting the chain branching. In this
sense, one can speak of ETC existence as functional (and, perhaps, even morphologi-
cal) subunits of the mitochondrial membrane. It does not mean that ETC do not interact with each other, i.e., that the functioning of one ETC does not depend on the
states of other ETC in the membrane. Moreover, we shall see in Chap.8 that membranes
containing ETC indeed have such cooperative properties, caused by their conformational changes.

The second question, to a considerable extent related to the first one, is
whether individual lipoprotein polyenzyme complexes, a kind of ETC subunits, really
exist within ETC.

As was stated above, GREEN and co-workers succeeded in fragmentating mitochondrial membranes into oligoenzyme complexes of four types: complex I containing FP_I
and FeS proteins; complex II, FP_{II} and FeS protein; complex III, cytochromes b,
c_1 and FeS proteins; and complex IV, cytochrome oxidase [7.84]. Kinetic properties
of carriers in the complexes differ from those in mitochondria. At the same laboratory, the normal functioning of various carriers was shown to require the presence of a special, so-called, structure protein [7.101]. It should be noted that
even a weak coercion on mitochondria, resulting in the formation of the so-called
submitochondrial particles (SMP), able to perform controllable electron transfer
and oxidative phosphorylation, leads to changes in kinetic properties of certain
carriers—cytochromes b and ubiquinone [7.102]. Moreover, SMP differ not only in
their membrane structure (the membrane is turned inside out as compared with intact mitochondria), but in the structural characteristics of certain protein carriers, e.g., FP_{III} [7.103-105].

It being possible to isolate Green's complexes indicates only that carriers,
having close values of equilibrium redox potentials and being parts of the same
complex, are indeed neighbors in ETC, as was already supposed on the grounds of
kinetic measurements, inhibitor experiments, etc. The preexistence of morphologically marked out formations, corresponding to these complexes in ETC, is hardly
imaginable. SKULACHEV was probably right when he wrote: "in a sense, the whole
respiratory chain may be defined as a single enzyme, NADH oxidase, and individual
carriers as subunits of different degrees of aggregation, constituting one enzyme
of extremely complicated quaternary structure" [7.106].

Let us now discuss some thermodynamic characteristics of ETC. In current
scientific literature, such discussion is solely based on values of standard redox potentials E_m^0 (midpoint potentials) of ETC constituting carriers. The physical

meaning of these values is, however, not completely clear. Equilibrium E_m^0 values
of the extreme ETC components, $NAD^+/NADH$ and $1/2\ O_2/H_2O$, are -290 and +800 mV,
respectively. The measured potential values for intermediary carriers, indeed,
happen to lie in this interval. However, the comparison of E_m^0 values of isolated
carriers with those for the same carriers in intact mitochondria or SMP, measured
by the concentration ratio of their oxidized and reduced forms, reveals discrepancies. Thus, just the inclusion of carriers into ETC changes their redox properties. Moreover, for certain carriers, these properties depend on the ETC state,
the presence of inhibitors, ATP and ADP concentrations, decouplers [7.69,80,87,97,
107,108]. The measured E_m^0 values of carriers such as cytochromes a_3 and b_T may be
changed under the action of ATP or decouplers by 100-300 mV. The authors cited
above suggested that these E_m^0 changes are the evidence of these carriers functioning as energy transducers in the process of oxidative phosphorylation (see
later). I think, however, that E_m^0 values of carriers, measured by the concentration
ratio of their oxidized and reduced forms, can hardly have the classical physical-
chemical meaning. It is not only a matter of possible differences between ETC
stationary and equilibrium states, or between stationary and equilibrium values of
the ratio of reduced and oxidized forms for certain carriers (functioning ETC is
an open system). What is more important, the degree of *kinetic* deviation from the
equilibrium state (see Chap.4) may depend on the path along which the system has
reached a given quasi-stationary state, and, therefore, the measured redox potentials of individual carriers may not be thermodynamic quantities. The concentration ratio of reduced and oxidized forms in a given quasi-stationary state may
not be equal to the ratio between the effective rate constants of reduction and
oxidation for the same carriers, of necessity measured on the disturbed system.

Let us consider now the data concerning the rates of electron transfer between
the components of mitochondrial ETC. One can notice a considerable discrepancy
between the results obtained by different authors and even by the same authors
in different publications. This is quite understandable, because kinetic properties are extremely sensitive to the state of mitochondria, and it is practically
impossible to ensure identical conditions. Even the existing data can, nevertheless,
lead us to certain important conclusions. Let us cite, for example, a kinetic
scheme of ETC from the studies performed at CHANCE's laboratory. In Fig.7.6 an electron transfer scheme of mitochondrial ETC, as described in [7.109], is shown, and
the oxidation rates of totally reduced carriers in anaerobic mitochondria after
oxygen addition by pulse or stopped flow methods are indicated.

$$O_2 \xleftarrow{0,40} a_3 \xleftarrow{0,51} a \xleftarrow{2,5} c \xleftarrow{5,0} c_1 \xleftarrow{80} b \xleftarrow{300} FP_{III} \xleftarrow{500} FP_I \xleftarrow{750} NADH$$

Fig. 7.6. Kinetic scheme of ETC carrier oxidation in mitochondria. Numbers over
arrows denote oxidation halftimes (in ms) after O_2 addition

$$c \xleftarrow{5} c_1 \xleftarrow{1500} b_T \xleftarrow{100} b_K \xleftarrow{1200} FP_{III} \qquad \text{(a)}$$

$$c \xleftarrow{5} c_1 \xleftarrow{100} b_T \xleftarrow{100} b_K \xleftarrow{300} FP_{III} \qquad \text{(b)}$$

Fig.7.7a,b. Kinetic scheme of ETC carrier oxidation in coupled (a) and uncoupled (b) mitochondria

More recent data, obtained in the same laboratory [7.110], are shown in Fig.7.7. Here we already see two cytochromes b : b_T and b_k, and the results of measurements on uncoupled ETC (free oxidation) and on phosphorylating ETC (coupled oxidation). Numbers above the arrows designate here halftimes (in ms) of electron transfer between adjacent carriers, initiated in "anaerobic" mitochondria by oxygen addition (and not the halftimes of oxidation, measured from the moment of oxygen addition).

Some remarks concerning these data are called for. First of all, the fastest reaction (between O_2 and cytochrome a_3) is a bimolecular process. According to the results, obtained later in the same laboratory [7.97] the second-order rate constant for reaction O_2 + cyt a_3 equals $3 \cdot 10^7$ mol^{-1} s^{-1} (for pigeon heart muscle mitochondria, at O_2 concentrations between 4 and 48 µmol. The transformation halftime, at O_2 concentration of 25 µmol and 21°C, was ~$5 \cdot 10^{-4}$ s. The rate of this reaction changes (diminishes) slightly for "uncoupled" mitochondria (as compared with the "coupled" ones). The presentation of rate constant values for other reactions of electron transfer in ETC is meaningless. Nominally, in concrete experiments, e.g., when the process is initiated by O_2 pulse addition to anaerobic mitochondria (or by photodissociation of the complex between cytochrome a_3 and CO in oxygen atmosphere), the oxidation of individual carriers obeys satisfactorily the laws of first-order kinetics. Therefore, some authors often calculate first-order rate constants by dividing ln 2 by transformation halftime, as it is usually done in the kinetics of gas-phase or solution reactions. It should be born in mind, however, that in the case of reactions taking place not in a statistical ensemble or reagents, but between fixed centers localized by necessity in one and the same ETC, the process rate depends not only on the total quantity of the centers of a given type, but on the state of adjacent centers just in those ETC where the given reactive centers are localized. In turn, the reaction rate depends not only on the ETC state, but on the path along which ETC has reached this state, i.e., on the system history. The calculated rate constants can, therefore, in this case, have the same physical meaning as in conventional kinetics provided an immense multitude of conditions is fulfilled. It is better to use directly measurable quantities (as, e.g., reaction half-periods), which do not claim to have any theoretical meaning.

Electron transfer from cytochrome a to cytochrome a_3 proceeds by about one order of magnitude slower than in the reverse direction (it is true for undamaged mitochondria; in isolated cytochrome oxidase these rates differ not more than by a factor of two).

It should be noted that there is a sharp dependence of transfer rates in certain ETC regions, on the phosphorylation coupled to electron transfer (incidentally, the comparison between Figs.7.7 and 7.6 leads to the conclusion that the data in Fig.7.6 relate to uncoupled mitochondria). A detailed study of cytochrome b_T oxidation has shown that it proceeds in a rather complicated way, and, at certain conditions, the b_T oxidation taking place after O_2 addition to anaerobic mitochondria may be preceded by its temporary supplementary reduction [7.69].

The comparison of data shown in Figs.7.6 and 7.7 with other facts concerning ETC and individual carrier functioning leads to some contradictions, and it is far from clear how these are to be resolved. It is easy to see that the rate of ETC electron transfer processes is greatly enhanced after cytochrome C has been passed on the way from NADH (or succinate) to oxygen (especially, in the case of coupled mitochondria). It means that in a steady state of mitochondria, when oxygen is in excess and does not limit the overall rate, the oxydation of cytochromes must proceed much faster (by about two orders of magnitude) than their reduction, and, therefore, cytochromes must be, practically, completely oxidized. The same conclusion was drawn by CHANCE, who compared the rate of cytochrome C oxidation with its turnover number in a steady state of respiratory control (see Chap.8) of phosphorylating mitochondria [7.111]. The turnover number equals only 5 s^{-1}, although, in a respiratory control steady state, a considerable fraction of cytochrome C (about 15%) is reduced. It probably means that the rates of electron transfer between the cytochrome system components in a steady state of ETC differ from those in experimental conditions shown in Figs.7.6 and 7.7, i.e., at oxygen interaction with anaerobic mitochondria.

As regards cytochrome C, one must note that a contradiction also arises, if the results shown in Figs.7.6 and 7.7 are compared with the characteristics of isolated cytochrome C redox transformations. We saw in Chaps.5 and 6 that the reduction and oxidation of cytochrome C are accompanied by essential structural changes. PECHT and FARAGGI [7.112] showed that, in the course of cytochrome C reduction in solution, conformational changes necessary to reach an equilibrium state of ferrocytochrome C, last hundreds of milliseconds. At the same time, the data presented in this section indicate that, after oxygen addition to anaerobic mitochondria, cytochrome C is oxidized (electron transfer cyt c — cyt a) and reduced (electron transfer cyt c_1 — cyt c) in a few milliseconds. This contradiction can be eliminated with the help of one of two possible assumptions. Firstly, it is possible that the conformational relaxation of a cytochrome C molecule within the mitochondrial membrane structure proceeds much more rapidly than in solution. From my point of view, this assumption is extremely improbable. More likely is the second assumption according to which, during electron transfer through cytochrome C to oxygen in the initially anaerobic mytochondria, in the first moments the conformational changes simply do not have enough time to be completed. Indeed, in anaerobic mitochondria, ferrocytochrome C exists in an equilibrium "reduced" conformation. After instantaneous oxygen addition, the active center of cytochrome C gives an electron and becomes oxidized in a few milliseconds, but the conformation of protein remains "reduced". Electron transfer from the next carrier (cytochrome C_1) is also realized in a few milliseconds. The lifetime of cytochrome C active center in the oxidized state during consecutive acts of electron transfer to and from cytochrome C is, therefore, not large enough for the conformational change to be realized. This situation must continue until the pool of fast carriers, localized on the "reducing" side of cytochrome C (i.e., cytochrome C_1 in Figs.7.6 and 7.7), is exhausted. Afterwards, the acts of reduction of the active centers in cytochrome C molecules will occur so seldom that molecules in the gaps will have enough time to be relaxed to an equilibrium "oxidized" conformation. We know that in solution this relaxation is accompanied by certain changes in spectral properties. The above assumption can, therefore, be verified. If true, it means that, to accept an electron from the neighboring ETC carriers, cytochrome C must not necessarily acquire an equilibrium "oxidized" conformation. Unfortunately, this assumption leads to the rejection of a very attractive idea of there being a special electron path in ferricytochrome C from the protein surface to the iron atom through parallel aromatic rings of tyrosines and tryptophan (see Chap.5). From my point of view, however, this idea is too good to be true. In Chap.8 we shall discuss in detail in what cases one can expect essential conformation changes of individual ETC carriers to be realized[4].

We now terminate, for the time being, the discussion of electron transfer processes in mitochondrial ETC, and turn to the analysis of photoinduced electron transport in chloroplasts and chromatophores of photosynthetic organisms. The mechanism of the primary act of photoinduced separation of charges will be the topic discussed in the following paragraphs. As a matter in fact, electron transfer between ETC components of chloroplasts and chromatophores is a dark process, just as the mitochondrial electron transport. Light absorption ensures only the formation of ETC components—electron donors and acceptors with properly situated electron levels.

Despite the great and ever-increasing number of publications concerning the electron transport photosynthetic systems, the exact ETC composition, as well as the sequence of carriers, and their kinetic properties, remain obscure and are the points at issue for many scientific schools. We shall not dwell at length on most of these questions. For the formulation of the problems discussed in this book, it is rather of little importance, for instance, whether a copper-containing protein — plastocyanine — is localized before or parallel to cytochrome f in the electron

---

[4]It should be mentioned that cytochrome C conformational changes in a membrane can proceed much slower than in solution. It was shown in a kinetic study of cytochrome C immobilized on various surfaces [7.113,201].

carrier chain between the active centers. At the same time, some specific properties of photosynthetic systems, associated with it being, in principle, possible
to initiate electron transfer processes by short-duration light flashes, did make
it possible to establish certain new important facts, which could not be up to
now reproduced with mitochondrial ETC. We shall dwell here, mainly, on these new
results, and begin with a rather schematic description of ETC structures in bacterial chromatophores, green plants and algae chloroplasts. In Figs.7.8 and 7.9
are, therefore, shown only those ETC components which are necessary for the understanding of the meaning of experiments that will be discussed later. Figure 7.8
represents the ETC scheme in chromatophores of sulphur bacteria, in which (during
noncyclic electron transport) electron transfer from donor—hydrosulphide ion
(HS$^-$) to acceptor— oxidized form of NADP (NADP$^+$) is realized. Particular attention
should be given to the bacteriochlorophyll molecule, P 890, situated in the photochemical active center. In its excited singlet electronic state this molecule is
a good electron donor (its formal redox pontial equals about -930 mV). Therefore,
absorption of the light quantum with a wavelength not exceeding 9000 Å leads to
charge separation, the electron being transferred to the primary acceptor, whose
nature is still unknown. According to some data, this acceptor is plastoquinone
(see, e.g. [7.114]), and to other an iron-sulphur protein of the ferredoxin (Fd)
type shown in Fig.7.8 [7.115].

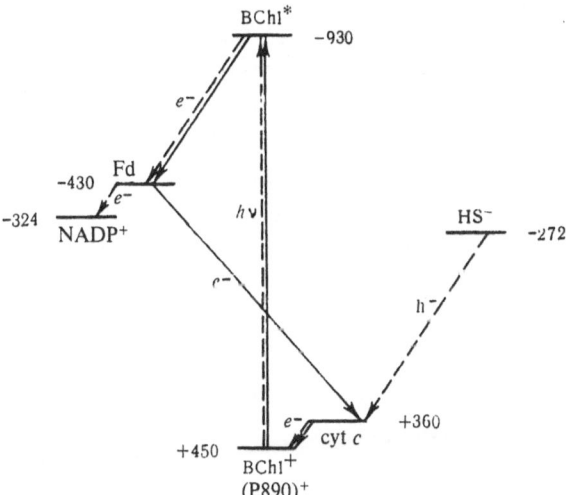

Fig. 7.8. Scheme of electron transport chain in chromatophores of sulphur bacteria.
Numbers denote equilibrium values of redox potentials (in mV) of corresponding
centers. Solid lines indicate the path of cyclic electron transport

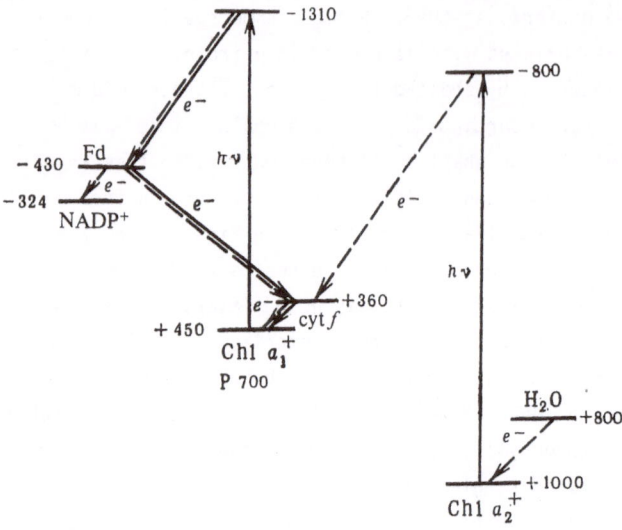

Fig. 7.9. Scheme of electron transport chain in chloroplasts of higher plants.
Numbers denote equilibrium values of redox potentials (in mV) of corresponding
centers. Solid lines indicate the path of cyclic electron transport

We shall return below to the question of the primary electron acceptor. Ferredoxin is, undoubtedly, situated in the ETC between P 890 and the final acceptor
$NADP^+$.

The positive center (P 890)$^+$, formed as a result of charge separation, is a good
oxidizing agent (its redox potential is about 450 mV), and accepts an electron from
the oxidized substrate (sulphide compounds) via the primary donor—cytochrome C
molecule, which, probably, forms a common protein complex with P 890. Thus, during
noncyclic electron transport, light quantum with energy of about 1.38 eV transfers
an electron from the substrate to the final acceptor expending only 0.05 eV. The
remaining energy of the quantum is spent on the ATP synthesis (photophosphorylation,
see below) and dissipates into heat. At certain conditions (excess of NADP, lack
of ATP), the electron flow from ferredoxin may go through a system of special carriers to the primary donor, cytochrome C, and again to (P 890)$^+$ (cyclic electron
transport). In the last case, electrons are moving along a closed cycle, and the
energy liberated at steps, corresponding to redox potential increase, is expended
on ATP synthesis only and dissipates into heat.

ETC of chloroplats of higher plants and green algae (see Fig.7.9) includes two
photosystems [7.51,116]^5. Photosystem I duplicates ETC of bacterial chromatophores,
with the difference that the photochemical active center contains, instead of bacteriochlorophyll P 890, chlorophyll a_1 having its long-wavelength absorption

---

^{5}New data concerning ETC of chloroplasts can be found in [7.202].

maximum at 7000 Å (P 700)[6], and the primary donor is cytochrome f, which, however, belongs to the cytochrome C type. In the case of noncyclic electron transport, however, oxidized cytochrome f obtains its electron not from the substrate (water) but from the primary acceptor of photosystem II through a system of carriers (not shown in the scheme). Chlorophyll a_2^+ — the oxidized active center of photosystem II — gets its electron from the H_2O molecule through another system of carriers. Thus, during noncyclic electron transport, the transfer of one electron from substrate H_2O to final acceptor ($NADP^+$) takes two light quanta with total energy of ~3.65 eV, out of which ~1.12 eV are used to ensure the transfer, the rest being dissipated into heat and spent on ATP synthesis[7].

It seems that the ETC of chloroplasts and chromatophores are not statistical, but (at least functionally, if not morphologically) isolated formations. Electron transfer between separate ETC is, probably, realized only through the plastoquinone pool (the plastoquinone molar concentration is by about one order of magnitude higher than that of other carriers) [7.118].

In a kinetic study of electron transfer in the ETC of chloroplasts, using light flashes of microsecond duration, we have recently shown the carrier chain between two photosystems to contain a two-electron "shutter" [7.119-121]. This "shutter" lets electrons pass through it only in pairs. The role of such a shutter seems to be played by plastoquinone molecules, transferring electrons through the hydrophobic layer of a thylakoid membrane. The work of this shutter is based on the differences in the solubility of plastoquinone in various redox states in hydrophobic and hydrophilic solvents. These experiments have also shown the interaction between plastoquinone molecules, located in different ETC, to proceed much slower than in the case of molecules in one and the same ETC.

We have already mentioned above that the possibility of photoinitiation of transfer reactions is very favorable for kinetic research. The most interesting are the results obtained in the course of investigating the transfer kinetics in definite ETC regions in wide temperature intervals. As early as 1960, it was shown in CHANCE's laboratory that cytochrome C oxidation in photosynthetic sulphur bacteria can proceed at liquid nitrogen temperatures [7.122]. The first experiments, concerning dark cytochrome f oxidation after photoinduced oxidation of P 700 in

---

[6]Although all wavelengths are given in Angströms, we use here notations P 700, P 890, etc., adopted in the literature.

[7]KNAFF and ARNON [7.117] have recently proposed a new scheme of chloroplast ETC. The main difference from the conventional scheme lies in the assumption, according to which ferredoxins, located in the chains of noncyclic and cyclic transports, are different molecules. The cyclic electron transport is ensured by separate photosystems with separate sets of carriers which do not coincide with the set of carriers of two light reactions ensuring the noncyclic transport. Irrespective of the correctness of this point of view, it does not introduce anything principally new into the physical aspects of the whole problem.

green plants, were published in 1961 [7.123]. CHANCE and BONNER [7.124] obtained kinetic evidence of the electron transfer from cytochrome f to (P 700)$^+$ in ETC of green leaves proceeding at 77 K at least as efficiently as at room temperature. Similar experiments were carried out with sulphur bacteria chromatophores [7.109, 125-131].

For one of the sulphur bacteric species, the following kinetic parameters of electron transfer from cytochrome C to BChl$^+$ after the photooxidation of the latter by laser flash were measured [7.127]. The reaction halftime value, $\tau_{\frac{1}{2}}$, at room temperature is 2.10^{-6} s. $\tau_{\frac{1}{2}}$ increases with temperature decrease, obeying formally the Arrhenius law with activation energy $E_a \approx 0.14$ eV. At 130 K, $\tau_{\frac{1}{2}} \approx 2.10^{-3}$ s. At 100 K, $\tau_{\frac{1}{2}} \approx 2.3 .10^{-3}$ s, and after that the halftime value remains practically constant down to 4 K ($E_a < 2.10^{-4}$ eV). The whole temperature interval between room and helium temperature can, thus, be divided into two parts: at temperatures above ~100 K, the electron transfer rate is temperature dependent, and below this value, temperature independent. Similar results were obtained by GRIGOROV [7.129] for other species of sulphur bacteria. The halftime value at room temperature was about $1.5.10^{-6}$ s. The rate of transfer drops till ~200 K ($E_a \approx 0.13$ eV) and, after that, remains practically constant down to liquid nitrogen temperature. Of great interest were the data concerning the temperature dependence of photooxidized cytochrome C dark reduction for the same species of sulphur bacteria [7.130]. Reduction proceeds in two stages, the first fast stage being, probably, caused by the transfer of electrons from photoreduced ETC carriers to cytochrome C. The rate of this fast stage has a complicated temperature dependence. With temperature decreasing from room temperature to 230 K, this stage slows down by about two orders of magnitude, then the rate increases again, and in the temperature interval of 170-77 K becomes practically constant. Comparatively slow dark reduction of photooxidized cytochrome C with a weak dependence on temperature in the interval of 295-77 K was also observed in the study of other species of sulphur bacteria [7.128].

It was also shown that some electron transfer reactions (oxidation of one of the cytochromes b and reduction of a new center, pigment P 546) in photosystem II of chloroplasts and green leaves can proceed at liquid nitrogen temperature [7.132].

Analyzing their results concerning the photoinduced changes in ESR signal of ferredoxin contained in photosystem I of chloroplasts JANG and BLUMBERG [7.133] did come to the conclusion that these secondary electron transfer reactions may proceed at temperatures down to 1.5 K.

Thus, the adduced data on electron transfer reactions in ETC of mitochondria, chloroplasts, and chromatophores, indicate that electron transfer takes place between centers tightly and specifically fixed in a membrane structure. The transfer conditions depend, essentially, on the membrane state. In some cases (mainly, for the processes located close to active centers of photosynthetic systems), electron transfer reactions were recorded whose rates are practically temperature independent. These data should be borne in mind when the possible mechanisms of electron transport in membrane structures are discussed.

7.4 Electron Transfer Reactions and Semiconduction in Biological Systems

When some new field of physics develops, or an already existing field becomes more popular, publications always appear devoted to the extreme importance of this field for biology. This tendency is quite useful, because it leads, in the end, to a thorough analysis of new physical aspects of the structures and processes in biological systems. Unfortunately, amongst the ardent adherents, going along with every new current, there are, as a rule, many physicists who do not know the substance of biology and many biologists acquainted with physics only at the level of popular-science literature. Therefore, their works aften arouse the irritation of biologists as well as physicists. In order to ascertain whether the new infatuation has any rational kernel, one must overcome this irritation.

Some time ago ideas about the extremely important role of semiconduction phenomena in biology were rather popular. In 1938 the well-known physicist JORDAN [7.134] suggested that many important biological phenomena can be explained, if it is postulated that giant biopolymer molecules have properties of solids, e.g., that they have energy bands formed due to the interaction of regularly spaced groups, and, therefore, energy migration over great distances is possible. On the basis of this assumption, the action spectrum of urease photoinactivation was discussed, and some ideas concerning mutation mechanisms and gene interactions were put forward.

In 1941 this idea was taken up by biologist SZENT-GYORGYI [7.135,136], and he was the first to assume that protein structures can also have properties of semiconductors, i.e., possess conduction bands along which electron or hole charges can migrate.

As regards the migration of electronic excitation energy in protein molecules and their complexes, numerous experimental data obtained during many years in different laboratories leave no doubt of its existence [7.137-141]. The mechanism of this energy migration in protein systems seems now to be clear: in protein macromolecules, the electronic excitation energy transfer between aromatic amino acid residues or that involving nonprotein chromophore groups (adsorbed dyes, prostetic groups, etc.) is realized according to the VAVILOV-FÖRSTER law of nonradiative resonance transfer [7.142,143]. This mechanism differs from that of radiative transfer. In the latter case, the transfer is determined by the electric field component in the energy flux emitted by the molecular oscillator, whereas resonance transfer does not involve any irradiation or absorption of quanta. Excitation energy migration in pigment systems of photosynthetic organisms seems to be realized in a slightly different way (see below).

Let us now discuss the possibility of a semiconduction mechanism of charge migration in protein structures. SZENT-GYORGYI's idea of the importance of this mechanism in principal biological processes was subjected to criticism [7.144,145]. The most essential objection was that the excitation of an electron to the conduction band requires, in the case of such systems as protein macromolecules, an energy of about 3 eV. It means that, in the usual temperature region, the conduction band must remain empty.

Later, the critics of Szent-Gyorgyi's concept used as their chief argument the results of calculations carried out by EVANS and GERGELY [7.146]. Having applied the conventional method of molecular orbitals (MO), these authors calculated the positions of molecular electronic levels arising as a result of π-electron interaction between peptide groups through hydrogen bonds $>N-H \ldots O = C<$ across polypeptide chains. By means of extremely approximate treatment they showed π conjugation without σ bonds through hydrogen bridges to lead to the appearance of very narrow ($\delta E < 0.2$ eV) energy bands, instead of the discrete MO levels of separate peptide groups. The distance from the lowest empty level (band) to the highest occupied level (band) decreases, as a result of interaction, from 4.46 eV for a separate peptide group to 3.05 eV for an infinitely long polypeptide chain. The authors concluded that protein macromolecules in the ground state have narrow conduction bands directed across the main polypeptide chain. These bands are lying too high to provide for the semiconduction properties of proteins in ground states, but can lead to protein photoconductivity. The narrowness of the conduction bands explains, according to the authors, the functional specificity of the protein.

This work stimulated a series of important studies, but it is difficult to understand now how such primitive and rather trivial calculations could at one time be regarded as proof of the existence or the absence of semiconduction properties in proteins. As a matter of fact, in this work, with the help of the MO method, the neutral excited states of one-dimensional molecular crystals consisting of weakly interacting peptide groups were calculated. The qualitative conclusions obtained by the authors could be reached without any calculations. The absence of low-lying levels of electronic neutral excitations of peptide bonds is well known from the spectra of simple peptides. The assumption of a weak interaction between the peptide groups must automatically lead to the appearance of narrow bands of slightly decreased energy. One can arrive at the same conclusion by comparing the optical characteristics of dipeptides and proteins.

In the conventional MO method without directly taking into account the interactions between electrons and using wave functions without antisymmetrization, one does not generally know whether the singlet or the triplet electronic excited state is being calculated. According to some considerations [7.147], the numerical values obtained are to be compared with the position of the triplet excited levels. In any case, the calculations concern the neutral excited states of a system, while the position of the conduction band in molecular crystals is determined not by the position of a neutral but that of a polar branch of excitation [7.148-153]. Moreover, even if we assume the coincidence of neutral and polar excited levels (i.e., of the neutral exciton and the transfer exciton levels), the treatment results are related only to the intrinsic semiconductivity. The real protein structures, however, contain an immense number of impurity centers (prostetic groups, metal ions, ionized side groups and just the aromatic amino acid residues). Thus, the calculations of EVANS and GERGELY, even if we accept that they give the position of conduction band, are, at best, the evidence of the absence of intrinsic semiconductivity in proteins, but not that of dark semiconductivity in general. On the strength of such considerations and of some experimental data concerning ESR spectra of biological objects (see below), I came in 1957 to the conclusion that there exists dark impurity semiconductivity in protein macromolecules and that this property plays an important role in redox processes [7.7]. I now think that these conclusions were erroneous. In actual protein structures with their multitude of electron and hole traps located (in space) close to the conduction band, the motion of charge carriers over any large distances, following the usual zone mechanism, is impossible.

Moreover, a detailed study of biological structures (membranes of mitochondria and chloroplasts) in which long-range dark transfer of electrons and holes is taking place, has shown the semiconducting properties to be simply unnecessary. In these systems, the charges are moving from one fixed center (the trap) to another, lingering on some of them long enough for the complicated chemical transformations in the closest vicinity of the trap to reach completion. The distances between the centers in these ETC can, probably, range from several angströms to several tens of angströms. It makes the semiconductivity ideas meaningless. The problem of possible mechanisms of such transfer is of great interest and will be discussed in detail in Sect.7.5.

Experimental studies of conduction properties of proteins and of more complex biological structures are, nevertheless, continued up to the present in many laboratories. Very often, the measurements are made on dried films and powders (tablets) of proteins and even of such complex structures as chloroplasts. These measurements are often made using direct current or low frequencies [7.154-158]. The results of these measurements are scarcely of any biophysical importance. The fact that all the samples studied have shown exponential temperature dependence of conductivity means nothing: this dependence is typical for all organic materials, even for manifest dielectrics. The conductivity of these objects may be determined by different mechanisms: processes in the vicinity of electrodes, ion migration, charge transfer between regular regions of high conductivity, etc. These mechanisms were discussed in detail in special monographs [7.153,159-161].

First results of a study of dielectric characteristics of biological objects in the 10^{10} cps frequency region have been recently published [7.162-169]. The investigators actually measured the real (ε') and the imaginary (ε'') parts of the dielectric constant. The idea of these experiments lies in the fact, that, at such high frequencies, intermolecular barriers cannot play an essential role, and measurements give directly the polarizability determined by free electron and hole charges. Unfortunately, no consistent physical theory of observed phenomena exists so far, even for the purest case of photoinduced changes in dielectric properties ("microwave photoconductivity"). More or less substantiated is, probably, the statement that orientational polarization cannot give any substantial contribution to the observed signals [7.170]. The effects observed in complicated biological structures seem to be caused, mainly, by the dark secondary processes (macrostructural polarization, ion displacements, etc.) [7.171].

In 1957 I suggested that a narrow singlet ESR signal in cells and tissues proves the existence of delocalized free electrons (or holes) and, therefore, can be regarded as a proof of semiconducting properties of these objects [7.7]. The data concerning ESR spectra of γ-irradiated native and denatured proteins were interpreted in the same way [7.8].

It is now quite clear that these views were erroneous (see Sect.7.2 and [7.172-175]).

In complex biopolymer intracellular structures, there, certainly, exist many centers, having donor and acceptor properties, which take part in electron transfer processes, and are, in principle, able to play the role of impurity centers. The realization of semiconduction mechanisms in the main biological processes is, however, practically impossible.

Probably, the only exception is the primary light process in photosynthetic systems. In pigment systems of higher plant chloroplasts and bacterial chromatophores, the bulk of the chlorophyll is functioning as a light-collecting antenna. The excitation migrates through the ensemble of packed dye molecules, and the photochemical act — the formation of an oxidized active center and a reduced primary donor — is realized on active centers, whose concentration is low as compared with that of the pigment (see Sect.7.3).

Recent experimental results show that a light-collecting pigment system represents an ensemble of regularly arranged chlorophyll molecules [7.176,177]. Energy migration within this ensemble is, probably, realized not by the resonance, but by the exciton mechanism. If so, it indicates a comparatively strong intermolecular interaction [7.178]. In this case, two different physical mechanisms of processes preceding the first photochemical reaction are, in principle, possible. 1) Charge separation takes place only on the active center, and the primary oxidizer [e.g., (P 700)$^+$], as well as the primary reductant (e.g., reduced ferredoxin), are formed during one act. It is only the neutral excitation that migrates through the pigment system molecules. 2) Semiconduction mechanism. Charge separation leading to the formation of free carriers or of transfer exciton [7.148,150], takes place not on the active center, but on other structural defects. The photochemical active center (e.g., P 700) either traps a hole (which has previously migrated through the pigment system), or realizes the dissociation of the transfer exciton, the hole being trapped by the active center, and the electron migrating to the removed primary acceptor. In fact, the concept of semiconduction is based on the assumption that photoinduced charge carriers appear in the pigment system. The real physical problem is to prove or refute this assumption.

At one time, an essential argument against the semiconduction mechanisms being responsible for the primary stages of photosynthesis was considered to be the fact that too low photoeffect quantum yields are observed in chlorophyll films [7.179, 180]. However, later measurements of the photoconductivity of chlorophyll films, carried out in the conditions of continuous and flash illumination [7.181,182], showed the previous results to have been erroneous. At 7050 Å (the maximum photosensitivity), the lower limit of photoeffect yield is not less than 15-20%, and free charge carriers appear not later than 10^{-8} s after the illumination has been actuated. This means that the formation of charge carriers represents one of the main ways of the transformation of energy absorbed by chlorophyll, and that triplet excitations can scarcely play the role of intermediates.

These data, naturally, remove only one of the objections, but cannot be considered as proof of the large yield of free carrier formation in the pigment systems of photosynthetic organisms. If, however, we agree that charge separation can take place in a pigment matrix, and that, consequently, the chlorophyll molecules themselves may be functioning as primary acceptors, an obvious question arises: why do we observe only positively charged paramagnetic centers (P 700^+) in the ESR spectra? Where are the signals from the radical-anion states? (About different types of photoinduced ESR signals in photosynthetic systems see [7.183,184]).

A similar problem arose at one time in the course of ESR spectra studies for organic semiconductors of low conductivity (polymers with conjugated double bonds, dyes, charge transfer complexes). The ESR signals observed in these systems were shown to be caused by the radical-ion states arising as a result of charge separation and the trapping of electrons and holes by structural defects (which play the role of impurity centers) [7.152]. Only one type of signals was, however, invariably observed. The search for the second center in the case of polyphenylacetylene, one of the typical polymeric organic semiconductors, led to the following result [7.185]: at low microwave power level ($<10^{-5}$ W) in addition to the usual ESR signal with $\Delta H_{1/2} \approx 10$ Gauss (signal "A"), there appears a second signal (signal "B") with $\Delta H_{1/2} \approx 16$ Gauss, whose relative integral intensity increases with the decrease of microwave power, and at $\sim 5 \cdot 10^{-8}$ W becomes equal to that of signal A. Signal B is one of the strongly saturating paramagnetic centers ($T_1 T_2$ values for centers A and B are $6 \cdot 10^{-11}$ and $3 \cdot 10^{-9}$ s^2, respectively[8]) (footnote see next page). Special experiments with the addition of electron donors and acceptors showed the usually observed signal A to be caused by the radical-cationic, and the easily saturating signal B by the radical-anionic paramagnetic centers. Similar results were later obtained for other organic semiconductors, dyes included [7.186].

Investigations of this kind have not been carried out up to now with photosynthetic systems. They may be rather useful for the purpose of clarifying the mechanisms of primary acts of charge separation and the nature of the primary electron acceptor in photosynthesis.

It is, of course, not excluded that the pigment matrix properties in photosynthetic systems do not resemble those of the organic semiconductors studied, and the arising radical-cationic and radical-anionic centers do not appreciable differ as to their ESP characteristics (saturation factors included). In this case, a contribution to signal I (photoinduced ESR signal of chloroplasts in photosystem 1) would be made not only by $(P\ 700)^+$ centers, but also by the radical-anionic states of the primary acceptor (if we assume that these acceptors are chlorophyll molecules or their ensembles). Therefore, the second question, that has not been answered so far, is: how many types of paramagnetic centers contribute to the usually observed signal I?

In 1963 BEINERT and KOK [7.187] carried out a detailed comparison between the number of P 700 centers in photosynthetic systems measured by ESR and in those measured by optical spectroscopy. They found the number of paramagnetic centers, contributing to ESR signal I, to be always 2-5 times greater than the number of photooxidized P 700 centers[9]. It is probable that, in addition to optically identified centers $(P\ 700)^+$, other paramagnetic centers, e.g., plastosemiquinone radical anions, also contribute to ESR signal I, although the g factor of semiquinone ESR signal must be displaced relative to the observed value (as, e.g., the g factor of signal III in [7.184]).

There are numerous indirect data to evidence the complicated nature of photoinduced ESR signal I. KAFALIEVA et al. [7.189] tried to distinguish the photoinduced paramagnetic centers, responsible for signal characteristics. They recorded dispersion signals in the conditions of fast passage at different values of high-frequency modulation amplitude (H_m). With the increasing H_m, the amplitude of the bell-shape dispersion signal passes through a maximum, whose position depends on the T_1 and T_2 values of the paramagnetic center. At higher H_m values, the dispersion signal splits into two signals, the characteristics of this splitting also being determined by the T_1 and T_2 values. The study of signal I with intensity

---

[8]The product of longitudinal (T_1) and transverse (T_2) relaxation times of paramagnetic centers determines their microwave power saturation.

[9]In a more recent work coincidence of spectroscopic and radiospectroscopic data was reported [7.188].

corresponding to about $5 \cdot 10^{16}$ spins $\cdot$ cm^{-3} at 77 K showed at least four types of paramagnetic centers, having different relaxational properties, to contribute to this signal.

The nature and origin of all photoinduced paramagnetic centers responsible for ESR signal I in photosynthetic systems is, thus, far from being clear. New quantitative, especially kinetic, measurements, are necessary.

It is also clear that the problem of the possible participation of semiconduction mechanisms in primary stages of photoinduced charge separation in photosynthetic systems (probably, it is the only important biological process where we can suspect the existence of this possibility) remains open. Here we must mention works concerning the thermoluminescence of photosynthetic objects [7.190,191]. Results of these works can be interpreted in favor of semiconductivity, but cannot be regarded as final proof.

7.5 On the Tunnelling Mechanisms of Electron Transfer Between the ETC Components

In Sect.7.3 we presented the experimental data indicating that the dark electron transfer between some neighboring ETC components in chloroplasts and chromatophores proceeds with a measurable rate even at 4 K. This rate does not depend on temperature within a rather broad interval of low temperatures. These results, naturally, suggest that there is quantum mechanical tunnelling between the neighboring carriers in ETC. Such a suggestion was, probably, first made by DE VAULT and CHANCE [7.126] on the basis of their data concerning the constant rate of electron transfer from cytochrome to bacteriochlorophyll in the temperature range between 130 and 30 K (later, down to 4 K). The reaction rate is, however, temperature dependent at temperatures higher than 130 K.

The authors believed this temperature dependence (assuming invariable barrier parameters) to exclude the tunnelling mechanism, and tried to explain it by the heat expansion of a potential barrier. This attempt failed, because their estimations led to a completely absurd value for the barrier height (4 keV). Therefore, DE VAULT and CHANCE explained the experimentally observed temperature curve as being the result of the superposition of two processes: usual over-barrier transfer and tunnel transfer, which is temperature independent and determining the overall process kinetics at low temperatures. This interpretation also seems to be improbable (see [7.129]). In this case, the barrier height would be equal to electron transfer activation energy in the high temperature region, i.e., would be as low as 0.14 eV. In order to coordinate the measurements at low and high temperatures, one must assume the value of 75 Å for the barrier width. The diameter of the cytochrome molecule does not exceed 36 Å. The existence of a barrier so low and broad seems, therefore, to be at least doubtful.

In 1967 CHANCE et al. [7.127] suggested a new explanation of these experimental facts. They assumed that the process of cytochrome oxidation with electron transfer to the photooxidized active center (bacteriochlorophyll cation), in the whole temperature range studied, proceeds according to the tunnelling mechanism, but with different levels at thermal equilibrium participating (they postulated, for instance, the existence of one excited tunnelling level at 0.14 eV above the ground level).

In 1968 GUTMANN [7.192] showed that, at a reasonable barrier height (1 eV) and width (21-28 Å), tunnelling ensures a faster electron transfer than the over-barrier mechanism even at room and higher temperatures. A detailed theoretical investigation of the possibility of tunnelling electron transport was carried out by GRIGOROV and CHERNAVSKY [7.129,131]. Our subsequent account is based on these materials (see, also [7.193]), and is qualitative. More strict quantitative estimations can be found in the literature cited above.

It should be mentioned that the concept of electron tunnelling transfer is explicit in the quasi-classical approximation only, i.e., in the cases when the barrier is sufficiently high and depends rather slightly on the presence or absence of the transferred electron. This was stressed in the critical statement of OSTERHOFF and KUPPERMANN [7.194] during the discussion following CHANCE's report. For ETC electron transport this condition is, however, fulfilled. The centers, between which the electron transfer is taking place, are sufficiently deep electron traps (not less than 1 eV), separated from each other by many atoms and atomic groups which contain plenty of electrons at all available local energy levels. Thus, the presence or the absence of one transferred electron does not, practically, affect the parameters of the barrier and the traps. Interaction between traps (carrier active centers) is also negligible: it is well known that electronic absorption spectra of carriers do not directly depend on their neighbor's electronic states. To a zero approximation we may, therefore, assume that, in each trap, the transferred electron is moving in a self-consistent field of nuclei and other electrons, and may approximate this field by a stable (the vibrations of atomic groups and the conformational changes being neglected) potential well, and consider the energy levels of only one additional (coming or leaving) electron. We can, moreover, confine ourselves to the consideration of the well ground electron level only, because the excited electron levels of the carrier active center, as a rule, lie far above the ground level. It is, therefore, quite legitimate to speak of electron localization, i.e., to regard the localized (in one of the wells) electron state as stationary and the presence of another well as a weak perturbation. One can also use such concepts as "the number of collisions per unit time" (of the electron with the barrier wall), transparency of the barrier, etc.

Let us, thus, consider the problem of electron transfer between two neighboring ETC components with the properties indicated above, i.e., the problem of electron transfer between the ground electron levels of two potential wells: left (L) and right (R).

Let us assume that $E_L^0 > E_R^0$ [10], and the direction of electron transport is from left to right (at the initial time moment, the electron is localized in the left well, the left carrier is reduced, and the right well is empty, the right carrier is oxidized). The well-known expression for the frequency of tunnel transfer (in the case of a square-topped barrier)

$$\omega = \omega_0 \exp\left[-\frac{2L}{\hbar} \sqrt{2m(u - E^0)}\right] \quad , \tag{7.4}$$

where ω_0 is the "oscillator frequency" of the electron in the well, L the barrier width, u the barrier height (from the bottom of the well), E^0 the electron energy level, and m the electron mass, cannot be used here because it is only valid in the cases when the spectrum of the right (empty) well is continuous. The general problem of tunnel electron transfer between wells having discrete spectra of the type shown in Fig.7.10 was analyzed by GRIGOROV [7.129].

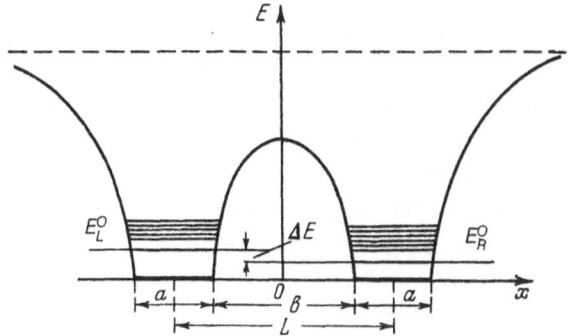

Fig.7.10. Approximation of ETC neighboring carriers by potential wells having discrete spectra [7.129]. Levels above E_L^0 and E_R^0 are vibrational sublevels of the ground electron state

In the case of discrete spectra, there arises the problem of energy balance, if $E_L^0 \neq E_R^0$. We have already mentioned that the left and the right wells do not, practically, interact. The term "practically" means, that the shift of E_L^0 and E_R^0, caused by interaction, is considerably smaller than the well depth. Level shift is maximal in the conditions of exact resonance, when $E_L^0 = E_R^0$. As a matter of fact, in this case we are dealing not with the shift, but with the resonance splitting of levels, ΔE_r

$$\Delta E_r = E^0 \exp\left[-\frac{L}{\hbar} \sqrt{2m(u - E^0)}\right] \quad . \tag{7.5}$$

If the difference between levels in the left and the right wells, $\Delta E = E_L - E_R$, is less than or equal to ΔE_r, no problem of energy balance arises. In this case,

[10]The upper zero index indicates pure electronic state.

quantum-mechanical oscillations occur. The electron that was initially in the left well passes within the time $\tau_r = \hbar/\Delta E_r$ to the right one, then, after the same time interval, returns, and so on [7.195]. The average electron density is uniformly distributed between the wells.

Introducing reasonable values for the parameters of ETC components into (7.5), we obtain $\Delta E_r \sim 10^{-6} - 10^{-4}$ eV. Therefore, to a zero approximation the resonance of different carrier levels seems to be very improbable (it must be recalled that condition $\Delta E = E_L^0 - E_R^0 \leq \Delta E_r$ must be satisfied in order to ensure the energy balance). When $\Delta E > \Delta E_r$ and the resonance requirements are not satisfied, tunnel transfer can only be realized if a certain process compensating the energy disbalance takes place.

Tunnelling can, in principle, be accompanied by induced molecular vibrations and other processes capable of taking up the energy. At first sight, it seems that, if all possible molecular vibrations have been taken into account, the energy spectrum must be almost continuous. Indeed, the density of vibrational levels, even in such a simple molecule as that of benzene, is about 10^4 eV^{-1} [7.196], and in more complex molecules the spectrum is even more dense. A vast majority of these levels are, however, combinatorial and almost each of the resulting vibrational states can be resolved into many normal vibrational states. Electron transfer can induce the excitation only of vibrational degrees of freedom that are strongly coupled with charge displacement. The transfer of one electron from the left to the right well leads to the change in equilibrium positions of the neighboring charged groups, and, consequently, in the frequencies of their vibrations. The "degree of coupling" between vibrational degree of freedom and electron transfer, to a first approximation, is proportional to the relative shift of equilibrium positions. The type of spectrum (its discrete or continuous character) is, therefore, completely determined by the groups in the immediate vicinity of an electron acceptor or donor (these are, in fact, the right and the left wells).

We have seen above that, in the systems of interest to us, ΔE_r is very small. We can, therefore, assume that, invariably, $\Delta E \gg \Delta E_r$. There are, in principle, three physical processes that can make electron tunnelling possible, if $\Delta E \gg \Delta E_r$: 1) birth of phonons; 2) energy transfer to a large number of normal vibrations strongly coupled with the electron; 3) energy transfer only to very few specific vibrations. This occurs in the case of extremely weak coupling of most vibrational degrees of freedom with electron displacement.

The problem of tunnel transitions, accompanied by birth and absorption of the phonons, was solved by GRIGOROV [7.129] with the help of the conventional methods of quantum theory of radiation. The following obvious result was obtained: this mechanism can play any essential role only when phonon wavelength is of the order of potential well dimensions. For the systems under consideration, $\Delta E \leq 10^{-3}$ eV is required, which corresponds to the already discussed version of the chance coincidence of levels.

The vibrational excitation process of type 2) is possible if the well environment contains a large number of rather mobile charges or polar groups as, for example, in the case of electron transfer between ions of variable valence in an electrolyte solution (see, e.g. [7.197-199]). This process, essentially, represents polarization of the medium and resembles the formation of a polaron. The situation here is analogous to the case of a continuous spectrum, and tunnelling can occur between the levels with an arbitrary energy gap. It is important to note that the excess energy, ΔE, dissipates (just as in the first case).

Electron traps of ETC components are inserted into lipoprotein membranes and are not in direct contact with the water medium of the cell. Therefore, the processes of type 2 can hardly play an important role in the biological electron transport. Probably, of particular importance are the processes of type 3, when the wells are protected from the ion medium by a nonpolar envelope and only few charged groups can be found near the site of electron localization. This can also be conceived as polarization. In this case, the relative number of strongly coupled vibrational degrees of freedom is small and, therefore, the vibrational spectrum is, practically, discrete.

There are two different mechanisms of energy transfer to vibrational degrees of freedom: (a) the excitation of vibration quantum in the right well by electron transfer; (b) the thermally induced change of vibrational frequency before electron transfer.

The probability of process (a) is temperature independent, and its value is determined by the coupling constant of electron displacement with normal vibrations. The probability of process (b) is proportional to that of excitation of a normal vibration quantum before electron transfer and is, therefore, temperature dependent. As a result, mechanism (a) is predominant at low temperatures, and mechanism (b) at high temperatures.

The condition of energy balance for process (a) can be written as

$$\Delta E = \hbar\omega_k = E_k \tag{7.6}$$

and for process (b) as

$$\Delta E = \Delta E_k = \hbar\Delta\omega_k \ . \tag{7.7}$$

Here, ω_k is the frequency of the excited vibrational quantum, and $\Delta\omega_k = \omega_k - \omega_k'$ is the difference between the "new" (ω_k) and the "old" (ω_k') vibrational frequencies. Spectroscopic investigations show that, for molecules similar to active centers of ETC carriers

$$E_k \approx \Delta E_k \approx 0.1 \ eV \ . \tag{7.8}$$

Normal vibrational quanta decay within $10^{-12} - 10^{-13}$ s, imparting their energy to phonons. There appears, accordingly, an uncertainty in the energies of these quanta of an order of magnitude $\Gamma \simeq 10^{-2} - 10^{-3}$ eV, and to satisfy the energy balance conditions, a strict fulfillment of requirements (7.6,7) is not obligatory. Otherwise the transfer probability would be proportional to $\delta(\Delta E, E_k)$ or $\delta(\Delta E, \Delta E_k)$. Factors of the type

$$\frac{\Gamma}{(\Delta E - E_k)^2 + \Gamma^2} \quad \text{or} \quad \frac{\Gamma}{(\Delta E - \Delta E_k)^2 + \Gamma^2} \tag{7.9}$$

must be used now instead of the δ functions. The above factors pass through a maximum for the case of resonance, i.e., when conditions (7.6,7) are satisfied.

We can, thus, expect tunnel electron transfer between neighboring ETC components to be effective, when the energy difference of their ground levels is about 0.1 ± 0.1 eV. In [7.129,131] it was shown that a theory based on these ideas can quite satisfactorily describe the known experimental data on the kinetics of electron transfer between neighboring ETC components at low and high temperatures. With a reasonable choice of ETC component parameters, the tunnel mechanism of electron transfer ensures a much higher transfer rate than the over-barrier mechanism. The last mechanism can, therefore, be neglected.

The acceptance of tunnelling as the basic mechanism of electron transfer in ETC of membrane structures in mitochondria and chloroplasts does not exclude the possibility of diffusive rapprochement of certain carriers to a distance at which tunnel transfer becomes actually possible.

The difference between the neighboring carriers in the energy of electronic levels must be close to the difference in their redox potentials. For many ETC regions, requirements (7.6-8) are fulfilled. Sometimes, however, this difference is considerably higher than 0.1 eV. For example, at the sites where the coupling between electron transfer and photophosphorylation is realized (see Chap.8), ΔE must be higher than 0.5 eV. In this case, the tunnel (as well as over-barrier) transfer is of very small probability and cannot ensure the observed rates of electron transport. One should remember, however, that the values of redox potentials apply to energies of a system in the equilibrium state. However, as we have seen in Chap.5, equilibrium conformations (or configurations) of a carrier macromolecule in the presence of an electron (the active center is reduced) and in its absence (the active center is oxidized) can differ essentially, and transition between the two conformations proceeds rather slowly. Thus, after electron transfer, a large part of the acceptor molecule appears to be in a nonequilibrium quasistationary conformation which slowly relaxes to an equilibrium one. In this case, the condition for tunnelling is not the coincidence (to an accuracy of ~0.1 eV) of electron levels of the levels of the reduced donor and the reduced acceptor in their equilibrium conformations, but the presence of an appropriately located electron level of the acceptor in the "oxidized" conformation. The energy released during tunnelling dissipates, but the energy slowly released by relaxation transition to a new equilibrium state can be used for the formation of macroerg (see Chap.8).

It should not be supposed that all problems arising in the analysis of physical mechanisms of electron transfer in biological systems have already been solved in principle. More or less accurate kinetic data in a broad temperature range are known only for few transfer processes in photosynthetic systems (mainly, in bacteria). The present-day theory is far from being perfect: it is essentially qualitative, and the quantitative estimations are too rough to be trusted.

We have already mentioned in Sect.7.3 the contradictory nature of many kinetic and thermodynamic experimental data concerning the functioning of ETC in mitochondria. In short, this branch of biological physics represents a gratifying sphere of action for scientists of different specialities, from theoretical physics to biological chemistry.

8. The Physics of Intracellular Energy Transformation and Accumulation

8.1 Overview

The problem of intracellular energy transformation and accumulation—one of the central biophysical problems—has many aspects. The primary energy sources for all organisms living on our earth are either sunlight (for plant and photosynthetic bacteria) or the energy liberated during chemical transformations (mainly, oxidation) of certain substances—food. This energy is transformed, accumulated, and then utilized to ensure a multitude of processes associated with vital activity. These are: synthesis of new low- and high-molecular compounds, creation of non-equilibrium distribution of metal ions and other low-molecular particles within heterogeneous tissues and intracellular structures, mechanical motion, luminescence, etc. After the pioneer work of LIPMANN [8.1] an idea began to spread and is almost generally accepted, according to which molecules of adenosine triphosphate (ATP) are functioning as a universal energy "keeper", as universal energy "taken money" in biological systems. Hydrolytic dissociation of the end phosphoric group from the ATP molecule is accompanied by a rather strong decrease of system energy. If this is true, the central question in the problem of energy transformation and accumulation in biology is then the question of the mechanism of ATP synthesis from ADP and P_i utilizing the above-mentioned energy sources. In some journals, however, a lively discussion has recently flared up concerning the question of whether ATP can be regarded as energy keeper. This discussion was often of terminological and rather naive nature, but in spite of (or, maybe, because of) it the analysis of this discussion may help us in formulating certain principal questions. In Sect.8.2, concerning the concept of macroergic bonds and the physical meaning of such statements as "energy is conserved in the form of macroergic ester bonds having heightened values of hydrolysis free energy", and "work is performed by utilizing the energy liberated during ATP hydrolysis", we shall consider the material of this discussion.

Having read some monographs and reviews concerning the mechanisms of ATP formation, one can come to the conclusion that the processes of substrate phosphorylation, i.e., those of ATP synthesis in the course of separate enzymatic reactions during glycolysis are, in principle, quite clear (although a few details are sometimes lacking), and the main puzzles are still the mechanisms of ATP formation in membrane structures, i.e., the processes of oxidative and membrane phosphorylation. One of the concepts of membrane phosphorylation, the so-called "chemical concept", suggests that the solution of the problem lies in the search for specific intermediate phosphorylated products in the fixed coupling sites, i.e., in reducing the problem of membrane phosphorylation to the "already solved and quite clear" problem of substrate phosphorylation. Careful consideration shows, however, that the physical principles of the coupling of energy-donating and energy-accepting reactions during substrate phosphorylation are as obscure as the corresponding principles for membrane phosphorylation.

Therefore, in Sect.8.3 we shall present the experimental data related to the problem of substrate phosphorylation, and discuss the possible physical mechanisms of this process. In Sect.8.4 we shall give the basic facts concerning ATP synthesis in the membranes of mitochondria and chloroplasts, and in Sect.8.5 we discuss the existing theories of this process. Finally, in Sect.8.6 a possible new approach to the problem of intracellular energy transformation and accumulation based on the ideas elaborated in the preceding chapters, will be formulated.

8.2 The ATP Problem

A detailed description of ATP and its role in the chemical transformations of various compounds in the course of intracellular bioenergetical processes can be found in all textbooks on biochemistry, in many monographs and reviews. Summing up, one can formulate the following generally accepted statements.

1) Intracellular chemical processes accompanied by the decrease of energy (usually they say: of free energy) ensure the ATP synthesis from ADP and ortho-phosphate

$$ADP + H_3PO_4 \rightarrow ATP + H_2O \quad . \tag{8.1}$$

2) Processes accompanied by energy increase are taking place due to ATP hydrolysis[1]

$$ATP + H_2O \rightarrow ADP + H_3PO_4 \quad . \tag{8.2}$$

There is no need to give examples proving these statements. They can be found in some comprehensive monographs (see, e.g. [8.2,3]) and in the selected original articles edited by KALCKAR [8.4]. Some remarks concerning (8.1,2) are called for.

---

[1]As a matter of fact, sometimes hydrolysis concerns not the end high-energy bond, but the high-energy bond whose hydrolysis leads to the formation of adenosinemono-phosphate (AMP) and pyrophosphate $ATP + H_2O \rightarrow AMP + H_4P_2O_7$. The existence of two ways of ATP hydrolysis is, probably, very important [8.5] but does not introduce any changes into the subsequent reasoning. We shall not, therefore, touch upon this question henceforth.

As is easily seen, ATP synthesis (8.1) is a reverse reaction of ATP hydrolysis (8.2). We have, nevertheless, written these reactions separately, but not in the form of one reversible reaction

$$ADP + H_3PO_4 \rightleftharpoons ATP + H_2O \ . \tag{8.3}$$

This has been done not because (8.1,2) are, in principle, irreversible. In each concrete case enzyme (or enzymes) accelerating reaction (8.1) accelerate reaction (8.2) to the same extent. Condensation (8.1) and hydrolysis (8.2) are, however, as a rule, localized at different sites and are catalized within the cell by different enzymes, at different local concentrations of ions essentially influencing the reaction rates. Therefore, there are many different reactions (8.1) and many different reactions (8.2) proceeding within the cell. For each of these reactions, its own equation (8.3) can be written. The arrows in (8.1,2) do not mean that these reactions are irreversible. They indicate that these reaction proceed under conditions when, in every concrete case, the initial state of the system (ATP, ADP, H_3PO_4 and H_2O) is shifted to the left of the equilibrium.

The second remark also concerns the conditionality of (8.1-3). Each component taking part in these reactions can exist in different states of ionization. Let us restrict ourselves, for the time being, to the thermodynamic aspects of this problem. In other words, let us temporarily forget the existence of various mechanisms caused by differences between enzyme catalyzing processes (8.1,2) and between the local conditions of these processes. We shall discuss a certain concrete equilibrium (8.3), whose position does not, in principle, depend on the presence of an enzyme[2]. In this case, (8.3) can have two meanings.

According to the first meaning ATP, ADP and H_3PO_4 represent corresponding neutral molecules, namely

ATP

ADP   H₃PO₄ .

---

[2]Provided, of course, the laws of equilibrium thermodynamics are valid. The problem of their applicability in this case will be considered below.

In this case, (8.3) has to do with one of many partial chemical equilibriums setting in the ATP, ADP and P_i system in water medium. The corresponding experimental equilibrium constant of hydrolysis reaction

$$K = \frac{a_{ADP} \cdot a_{H3PO4}}{a_{ATP}} \qquad (8.4)$$

is the true equilibrium constant of this partial reaction. In (8.4) a_i is the activity of ith compound in the water phase at equilibrium. The K constant is, naturally, associated with (8.3), for which the initial and end reaction products are dissolved in one and the same water medium (with one and the same content of salts), and do not depend on the reaction path. As is usually the case, a standard free energy

$$\Delta F^0 = -RT \ln K \qquad (8.5)$$

of the process is associated with K. ΔF^0 is a function of temperature and ionic strength.

According to the second meaning, the terms ADP, ATP and H_3PO_4 designate all ionization states of corresponding molecules (and sometimes of their M_g^{2+}, Ca^{2+}, etc., salts). Then the quantity

$$K_{obs} = \frac{a_{ADP} \cdot a_{H3PO4}}{a_{ATP}} \qquad (8.6)$$

is not a true equilibrium constant of a certain partial reaction, but can be in a complicated way expressed through the true constants of various partial reactions. The quantity

$$\Delta F^0_{obs} = -RT \ln K_{obs} \qquad (8.7)$$

is the change in the system free energy at unit summary activities of all the ionization states of ATP, ADP and P_i in the medium, associated with hydrolysis of 1 mole of ATP in all ionization states. The value of ΔF^0_{obs} depends on temperature, ionic strength, pH and the medium ionic composition.

In the present chapter we shall use (if not stated otherwise) the terms "process equilibrium constant" and "free energy of ATP hydrolysis" in the second sense, i.e., we shall consider only the effective observable quantities. As to their connection with the energetics of real intracellular processes, this will be clear from the subsequent discussion. It is, however, clear that if any equilibrium thermodynamic characteristics of (8.1,2) have a direct bearing on bioenergetics at all, they must be just these effective quantities.

Direct measurement of K_{obs} and ΔF^0_{obs} values is rather difficult, because, as a rule, equilibrium (8.3) is, practically, completely shifted to the left, and K_{obs} exceeds 10^5 M. Therefore, at any reasonable ADP and P_i concentrations the equilibrium is established at a negligibly small concentration of ATP. Usually,

K_{obs} and F_{obs}^0 values are obtained indirectly, by measuring the equilibrium constants of several reactions, their algebraic sum giving reaction (8.3). When performing these calculations one must assume that the equilibrium but not the stationary states are measured. It restricts the characteristic times of the processes in question (see Sect.3.1). Moreover, one must also assume that the mass action law is applicable to all reactions involved, and this assumption requires that proteins should be absent among the initial and the end reaction products (see Chap.5).

The most complete data can be found in [8.6,7]. In these studies the authors used the data on glutamine enzymatic synthesis from glutamate and ammonia with the participation of ATP, as well as the data on glutamine enzymatic hydrolysis

$$\text{Glutamate} + NH_3 + \text{ATP} \rightleftharpoons \text{Glutamine} + \text{ADP} + P_i \tag{8.8a}$$

$$\text{Glutamine} + H_2O \rightleftharpoons \text{Glutamate} + NH_3 \quad . \tag{8.8b}$$

Evidently,

$$K_{obs} = K_a \cdot K_b \quad ,$$

where K_a and K_b are the equilibrium constants of (8.8a,b), respectively.

ALBERTY [8.8,9], using previous results [8.10], drew contour charts for ΔF_{obs}^0 at ATP hydrolysis as a function of pH and pMg. SHIKAMA [8.6] extended these calculations to the systems with various Ca^{2+} concentrations.

ROSING and SLATER [8.7] revised earlier results. With the help of accurate direct measurements, they showed the K_a value used in all previous calculations to be ~5 times overestimated. They also corrected the stability constant value of the Mg^{2+}-glutamate complex, took account of the binding of K^+ ions, ATP, ADP and P_i, and introduced corrections into activity coefficients. The numerical values of thermodynamic parameters of reversible ATP hydrolysis, obtained by these authors, are, probably, the most reliable up to date.

At physiological conditions, ATP and ADP concentrations are almost equal, and the concentration of P_i is about 10^{-2} - 10^{-3} M. The true absolute ΔF value of the hydrolysis (or synthesis) or 1 mol of ATP within the cell must be, therefore, higher than the standard ΔF^0 values.

The absolute standard free energy values for the hydrolysis of the esters of orthophosphoric acid are, as a rule, lower than those for reaction (8.3). Thus, ΔF^0 for glucose-6-phosphate hydrolysis is as low as 3 kcal/mol. There exist, however, intermediate cases (see, e.g. [8.11]). Therefore, subdivision of organic phosphoric esters into "energy rich" and "energy poor" is to a certain extent a relative one. Nevertheless, the fact that ATP and ADP belong to the class of phosphoric esters with a high (compared with the majority of similar compounds) absolute value of hydrolysis free energy has a direct connection to ATP and ADP functioning. I, therefore, think, that the terms "macroerg" and "macroerg bond" (or "high energy bond") are quite justified. In the course of the above-mentioned discussion the concept of macroerg bonds (or compounds), as well as the notion of ATP as the intracellular energy keeper, were subjected to a violent criticism, mainly in the articles of BANKS and VERNON [8.11-15]. All critical statements were discussed in detail in [8.16-19]. In the course of this discussion, certain principal problems of bioenergetics were touched upon. We shall, therefore, use the materials of this discussion in our subsequent treatment of these problems.

First of all, we must analyze the concept of the high energy bond. In our analysis we shall use the terminology of conventional thermodynamics, i.e., the concepts of free energy and entropy. The question of the validity of these concepts and terminology in various concrete cases will be considered later. Already in their first article [8.12] GILLESPIE et al. asserted that the terms "macroerg bond" or "high energy bond" are meaningless because, firstly, we are dealing not with stronger but with weaker bonds (the reaction of hydrolysis is accompanied by a decrease and not by an increase of free energy), and, secondly ΔF^0—the standard change in free energy—refers to a process and not to a bond. This kind of reasoning is rather widespread in the literature (see, e.g. [8.20,21]). I think that this reasoning is erroneous. The concept "high energy compound" has the following physical meaning: for a given system in a given state (e.g., for ATP in water medium) there exists a kinetically available state that is characterized by a lower free energy level. Energy change, due to the transition between these states, is greater than for most of analogous processes. Beginning with Lipmann, nobody has attributed another meaning to this concept.

The statement that energy is stored in ATP in no way differs from the statement that energy is stored in an suspended weight and can be utilized in the course of its transfer to a lower kinetically available level, or from the statement that energy is stored in a $2H_2 + O_2$ mixture and can be utilized if we create conditions for the transition to the new kinetically available state $(2H_2O)$ with a lower-lying free energy level (for example, if we initiate the chain reaction of hydrogen combustion by temperature increase). The kinetic stability of ATP (in the absence of specific enzymes), of a weight (until there is a support under it), and of a hydrogen-oxygen mixture (until the temperature has reached the first combustion limit) does not bear any relation to the essence of the problem. To be sure, the attribution of system energy to one bond only ("macroerg bond") is arbitrary and has only the following meaning: during the transition from the state with a high free energy level to the state with a low free energy level only this bond is subjected to hydrolysis. No serious scientist has ever put another sense into this attribution. The concept of "bond energy" in chemistry is always more or less arbitrary. Thus, the energy change during the reaction of methyl radical formation from methane

$$CH_4 \rightarrow \overset{\bullet}{C}H_3 + \overset{\bullet}{H}$$

is often called "the energy of C-H bond" although in the course of this process
not only the two-electron C-H bond is ruptured but the whole residue of the molecule is rearranged with changes taking place in electron density distribution,
valence angles and interatomic distances.

One can, in principle, give a different definition of C-H bond energy in methane,
namely, define it as a quarter of the energy liberated during the synthesis of a
CH_4 molecule in gas phase by a hypothetical reaction between C and H atoms

$$C + 4H \rightarrow CH_4 \quad .$$

The C-H bond energy value will in this case, naturally, differ from that measured by the first method.

The terminological part of the criticism aimed at the concepts of modern bioenergetics presented in the papers cited above cannot, thus, be regarded as serious.
Any statement and any term can lead to a misunderstanding when a wrong meaning is
ascribed to them. Physicists use the term "force of electric current"[3], without
being embarassed by the fact that this "force" does not bear any relation whatsoever to the force in the Newtonian sense of the word.

Discussing the concept of high energy esterphosphate bonds one can ask a
question: why are the absolute values of free energy of hydrolysis for these bonds
higher than those for the majority of similar compounds. There are many papers in
the literature concerning this question (see[8.21-23]).

From my point of view, this problem cannot be considered very important. On the
one hand, it does not differ from an immense number of similar problems of structural chemistry, and, on the other hand, the differences of a few kcal/mol in the
transformation energies of the molecules of such complexity cannot be estimated
with certainty by means of the existing quantum mechanical methods for the treatment of multi-electron and multi-center systems. Consequently, the clarification
of the structural nature of macroerg bonds is, on the one hand, of no interest,
and, on the other, for the time being impossible.

In the criticism of the main tenets of bioenergetics the authors of [8.11-13]
spent much time on the following reasoning: reactions (8.1,2) do not proceed within
the cell "in actual reality", and one cannot infer, therefore, that the energy
liberated during some other process is conserved as a result of reaction (8.1)
completion, and that the energy liberated during reaction (8.2) ensures the completion of some endoergonic process (e.g., muscle contraction). According to this
criticism, the true mechanism of the overall process is, as a rule, unknown, and
the apportionment of stages (8.1,2) is entirely arbitrary. This is quite true but
does not bear any relation to the gist of our problem. Let us suppose that in a

---

[3]In Russian, the term "seela toka" ("the force of current") is used to designate
the strength of current.

multitude of cells 1 mole of glucose is oxidized

$$C_6H_{12}O_6 + 6O_2 \rightleftharpoons 6CO_2 + 6H_2O \quad . \tag{8.9}$$

This equation does not describe any elementary chemical reaction taking place within the cell in actual reality. It is a conventional formula which has the following meaning: a complex overall process of intracellular glucose oxidation, proceeding through an immense number of stages which involve different low- and high-molecular compounds, can be carried out in such a way that the disappearance of every glucose molecule will be accompanied by the disappearance of 6 oxygen molecules and the formation of 6 carbon dioxide and 6 water molecules. The water medium contained initially the first system: $C_6H_{12}O_6 + 6O_2$, and after the process has been completed, the second system: $6CO_2 + 6H_2O$. If the cells are in a steady state of growth they will not change with the reaction having been completed. The overall process may with a sufficient degree of accuracy be regarded as isothermic and isohoric. The maximum work which can be performed in the course of transformation (8.9) must be, therefore, equal to the differences between free energy values of the first and the second systems in water medium[4], i.e.,

$$\Delta F = F_1 - F_2 \quad .$$

We know that, for this process in water medium at standard conditions, $\Delta F^0 \approx 680$ kcal/mol.

At the same time, the oxidation of 1 mole of glucose within the cell can be performed in such a way that the sole overall chemical process [besides (8.9)] would be the formation of 36 moles of ATP from ADP and P_i according to (8.1). This formation must be accompanied by the free energy increase of about 324 kcal/mol. (If we assume that ΔF for the hydrolysis of 1 mole of ATP within the cell is equal to about 9 kcal/mol).

In this case, the process efficiency would be about 48%. We can say that energy conserved earlier in the system $C_6H_{12}O_6 + 6O_2$ has been partly transfered into the system ATP + H_2O. Only this part of energy, which did not undergo dissipation, can be, in principle, utilized to perform useful work.

In her paper [8.13] BANKS contended that, in most cases, it is meaningless to speak of some endergonic process being realized due to ATP hydrolysis. Among other processes, she discussed the following synthesis:

$$XOH + YOH + ATP \rightleftharpoons XOY + ADP + H_3PO_4 \quad . \tag{8.10}$$

This process is usually represented as a sum of two reactions

$$XOH + YOH \rightleftharpoons XOY + H_2O \tag{8.11}$$

---

[4]About the applicability of equilibrium thermodynamics in this case see below.

$$ATP + H_2O \rightleftharpoons ADP + H_3PO_4 \quad . \tag{8.12}$$

Biochemists usually say that endergonic reaction (8.11) proceeds at the expense of exergonic reaction (8.12). Banks, however, maintained that ATP hydrolysis does not proceed within the system "in actual reality", ATP playing, probably, the role of phosphorylating agent, i.e.,

$$XOH + ATP \rightleftharpoons XOP + ADP \quad . \tag{8.13}$$

The end product XOY is, probably, formed due to reaction

$$XOP + YOH \rightleftharpoons XOY + H_3PO_4 \quad . \tag{8.14}$$

Process (8.10) can, thus, be more properly represented as a sum of (8.13,14) than as a sum of (8.11,12). Why, asked BANKS, in our explanation of the process energetics, do we use (8.12), which does not proceed "in actual reality", if ATP is functioning properly?

This reasoning seems to be quite logical, but it is, nevertheless, quite wrong. First of all, (8.13,14) bear as little relation to the "actual reality" as (8.11, 12). We do not know in detail the mechanism of the overall process (8.10), but we do know that, in any case, this process includes a multitude of intermediate steps with the participation of enzyme molecules which are undergoing chemical and conformational transformations. However, as long as the approximation of equilibrium thermodynamics works, i.e., as long as free energy is a function of the system state and its change does not depend on the process path, the energetics of the overall reaction does not depend on its mechanism. It means that we have the right to subdivide (8.10) into any number of stages, provided their algebraic sum gives (8.10). It is quite natural to pick out the (8.11) stage: the overall process (8.10) leads to the condensation of residues X and Y. This condensation requires energy (in water medium, XOY hydrolysis is accompanied by a decrease of free energy). We see, however, that the overall process (8.10) is energetically favorable, because such choice of the first stage must necessarily lead to the complementary stage (8.12), corresponding to a considerable decrease of free energy. Thus, in the realm of equilibrium thermodynamics, there exists only one correct answer to the question: why can the energetically unfavorable condensation of X and Y into XOY be realized in the course of the overall process (8.10). It is: because the energetically favorable ATP hydrolysis serves, in this case, as an energetically favorable "complementary" (up to the overall process) stage. The choice of reversible ATP hydrolysis (i.e., reaction (8.12) as a compulsory stage in the thermodynamic analysis of many biochemical processes is justified because this reaction can be used as a universal and convenient part of quite different processes, from synthesis (8.10) to muscle contraction and active transport of ions. It is in this choice that the importance of the first work of LIPMANN [8.1] lies. He established the principles of classifying and scrutinizing from a common point of view the

immense multitude of intracellular reactions. LIPMANN has thus transformed the vast
chaos of separate facts into the well-proportioned edifice of modern biochemistry.
The objective prerequisite of this choice being successful was, of course, the fact,
that ATP takes part (i.e., is either an initial or an end reaction product) in the
majority of important biochemical processes.

The strong point of the thermodynamic approach, its generality and universality,
is, naturally, the reason of its limitation. Thermodynamic can only allow or forbid
some overall process, but cannot predict the kinetic possibility of its realization.
Thermodynamics cannot give the details of one or the other mechanism just because
this science is based on potentials which do not depend upon the paths of processes.

It is, however, quite sufficient, that the thermodynamic approach has allowed us
to formulate the principles of modern classification in biochemistry and to make
conclusions concerning the possibility or the impossibility of certain intracellular
reactions. Refusal to accept ATP hydrolysis (8.12) as a bioenergetical "standard"
is unreasonable not only because any imaginable standard is as far from (or as close
to) the actual reality as ATP, but because ATP participates in almost all the important biochemical processes.

Let us consider now the problem of the applicability of equilibrium thermodynamics to systems removed from the true thermodynamic equilibrium. The authors of
[8.11-13] contended that processes (8.9) or (8.12) with large formal values of
standard free energy changes never achieve equilibrium states, within the cell.
Thus, the "equilibrium constant" of glucose oxidation ($\Delta F^0 \approx 680$ kcal/mol) is

$$K = \frac{a_{CO_2}^6}{a_{O_2}^6 \cdot a_{C_6H_{12}O_6}^6} \approx 10^{495} \text{ mol}^{-1} \quad ,$$

where a_i is the activity of the ith component. It is clear that, at any imaginable
condition, the equilibrium can be established only after a practically complete
oxidation of glucose. Glucose oxidation, therefore, proceeds within the cell in the
conditions far removed not only from equilibrium, but beyond the limits where
Onsager's relations are valid. Such concepts of equilibrium thermodynamics as free
energy — a function determining the maximal useful work that can be performed by a
system — become meaningless.

This reasoning is wrong. If it were true, the laws of equilibrium thermodynamics
could not be applied, e.g., in the case of electrochemical processes, to the deduc-
tion of the Nernst formula, in particular. Indeed, in the Daniell cell, for example,
an overall chemical process

$$Zn + Cu^{2+} \rightleftharpoons Zn^{2+} + CU$$

is realized with $\Delta F^0 \approx 25$ kcal/mol and equilibrium constant $K \approx 2 \cdot 10^{18}$.

It means, that the true thermodynamic equilibrium will be established when the concentration of Zn^{2+} ions becomes 10^{18} times higher than that of Cu^{2+} ions, i.e., when the metallic zinc electrode has been completely dissolved, and all copper ions precipitate on the copper electrode in their metallic form. It is well known, however, that the properties of a galvanic cell can be successfully described by the formulas of equilibrium thermodynamics. The fact is that, in this case, the system is in a nonequilibrium state due to kinetic reasons: the concentrations of Cu^{2+} and Zn^{2+} ions change extremely slowly in comparison with the time intervals necessary to reach equilibrium with respect to all the other degrees of freedom (see, also, Chap.3). We can, therefore, treat all processes within this system as procceeding reversibly at fixed constant values of Zn^{2+} and Cu^{2+} concentrations, which play the roles of system parameters. The same can be said about the intracellular processes (8.9,12). Moreover, the existence of degrees of freedom relaxing but slowly to the equilibrium is a compulsory requirement for using the system motion along these degrees of freedom to perform external work.

Let us discuss one simple example: the expansion of ideal gas into vacuum. A cylinder subdivided into two volumes V_1 and V_2 by a partition is placed in a thermostat. The partition can be moved out, friction is neglected, the material of the walls is assumed to be incompressible. Volume V_1 initially contains a gas under pressure p_1 at thermal equilibrium ($p_1 V_1 = nRT$, where n is the number of gas moles). Volume V_2 is vacuous. Let us move the partition out, i.e., let the gas volume increase to $V_1 + V_2$. It is an isothermal process without any thermal exchange with the thermostat (the gas is ideal). In the new state the gas volume equals $V_1 + V_2$, and gas pressure $p_2 = p_1 V_1/(V_1 + V_2)$. The intrinsic energy of the system does not change, because the gas is assumed to be ideal, the temperature is constant, and the partition has been moved horizontally. The system entropy increases by $nR \ln(V_1 + V_2)/V_1$. Although the free energy of the system has been decreased by $nRT \ln(V_2 + V_2)/V_1$ (bound energy), no useful work has been done.

Let us now change the design of the device and carry out isothermal expansion in a different way. There is now a piston instead of a partition (this piston is assumed to be weightless, its friction with the wall is neglected). There is a weight M_1 on the piston, such that

$$gM_1/A = p_1 = nRT/V_1$$

where g is the acceleration of gravity, A the piston area. Let us change (instantly) M_1 for a smaller weight M_2 [$gM_2/A = p_2 = nRT/(V_1 + V_2)$]. The system is now out of equilibrium, the gas expands isothermally (now, due to thermal exchange with the thermostat), and the piston will ultimately get into a new equilibrium position. In this new state the gas will again have temperature T, volume $V_1 + V_2$, and pressure p_2. The system's intrinsic energy, however, increases in this case by $\Delta U = M_2 gh$, because the weight M_2 (which is a part of the system) will be raised to the height of h as a result of the process. If we do not include M_2 into the system, we can say that our system has performed external work M_2gh. The heat $Q = M_2gh$ obtained by the system from the thermostat has been used to do this work. We did not carry out this process in an optimum way, because the replacement of M_1 by M_2 has been made instantly. When the weight is diminished gradually the system will be able to receive more heat from the thermostat. In the case of infinitely slow weight decrease (from M_1 to M_2) the limit value of the heat received and the work performed will be equal to $nRT \ln(M_1/M_2) \equiv nRT \ln(V_1 + V_2)/V_1$.

What is the principal difference between these two processes? If we do not include the raised weight into our system the initial and the end states of the system are identical. In the first case, however, the decrease of system free energy has not been utilized. In the second case, the heat received has been used to perform external work (the raising of the weight). It was possible due to changes in the system construction. The possibility appeared to excite a degree of freedom which does not easily exchange energy with the other degrees of freedom of the system. Moreover, the motion along this specific degree of freedom is so slow that at every moment there is enough time for the equilibrium energy distribution among all the other degrees of freedom to be achieved.

This specific degree of freedom corresponds, in our case, to the mechanical motion of the piston and the weight along the cylinder axis. This mechanical degree of freedom does not, practically, interact with other degrees of freedom, and the energy (potential energy of the raised weight) can be conserved in the system indefinitely (until the piston or the weight are destroyed). The excitation mechanism of a specific degree of freedom is, in this case, quite clear. The possibility of excitation is determined by the system construction and by the fact that the piston and the weight are solid bodies with rigid bonds between atoms.

In all cases, when energy is stored in the form capable of performing useful work, we are dealing with excitation of specific but slowly relaxing degrees of freedom. In this sense, "molecular machines" do not differ from conventional machines[5]. Accumulation of the energy, liberated in the course of glucose oxidations in ATP, signifies the excitation of a slowly relaxing degree of freedom corresponding to the chemical transformation (8.3) (ATP molecule is stable, ATP + H_2O system is separated from ADP + H_3PO_4 system by a high kinetic barrier). The true problem lies in the excitation mechanism of slowly relaxing degrees of freedom. These mechanisms, which determine the processes of intracellular energy transformation, will be considered later.

Now, a few words about the use of terms "free energy" and "total energy" in the analysis of bioenergetical transformations. Almost all intracellular chemical

---

[5]These problems were discussed in detail by McCLARE [8.19,24]. A beautiful analysis of molecular machine functioning can also be found in [8.128-130].

processes are taking place within a heterogeneous medium. The low-molecular compounds or the active groups of biopolymers undergoing chemical transformations to a considerable degree lack the freedom of motion because they are more or less strictly fixed by macromolecular or membrane structures. Therefore, when concrete mechanisms and energetics are being discussed, individual macromolecules or macromolecular complexes must be regarded as statistical systems, and the conformations and configurations of these units as the states of these systems (as was done in Chap.4 in the analysis of the statistical thermodynamics of biopolymers). Energy transformation during elementary intracellular processes is, as a rule, realized as the system moves along specific slowly relaxing mechanical degrees of freedom. In these cases one must, therefore, use the concept of total energy of a system including the parametrically given energies of the interaction between the given system and the environment.

At the same time, considering overall chemical transformations when the products of initial and end reaction (or reactions) can be assumed to be in identical conditions, one must use the concepts and equations of equilibrium thermodynamics.

8.3 Substrate Phosphorylation

ATP can, thus, be regarded as the main intracellular energy accumulator. The physical problems of bioenergetics are, therefore, reduced to those of ATP synthesis mechanism. Almost all intracellular processes of ATP synthesis can be represented in the form

$$A_{ox} + B_{red} + ADP + P_i \rightarrow A_{red} + B_{ox} + ATP + H_2O \; . \tag{8.15}$$

It is necessary once again to emphasize that such a form is only a scheme of a "gross process", but not an equation of elementary chemical reactions proceeding in actual reality. It is customary to write (8.15) in the form of two coupled reactions [8.3].

$$
\begin{array}{ll}
A_{ox} + B_{red} & \longrightarrow ATP + H_2O \\
& \times \\
A_{red} + B_{ox} & \longleftarrow ADP + P_i \quad .
\end{array}
\tag{8.16}
$$

The meaning of this form is in the subdivision of process (8.15) into two processes, one of which, the redox process, is accompanied with a decrease, and the other, that of condensation, with an increase of system energy. This subdivision into stages presupposes that the redox reaction is coupled with one of the acid-base type, i.e., makes the latter thermodynamically possible. Scheme (8.16) is not entirely only a form of writing. In fact, the two processes in (8.16) can be decoupled, i.e., the redox process can be carried out without ATP formation. It is customary to assume [8.3] that an additional process of "primary macroerg" formation takes place, between the redox reaction and ATP synthesis. This ensures the coupling of the two stages (8.16). It can be written out in the form

$$A_{ox} + B_{red} \rightarrow X\sim \rightarrow ATP + H_2O$$
$$A_{red} + B_{ox} \rightarrow X \rightarrow ADP + P_i \qquad . \tag{8.17}$$

Here, X can designate a chemical compound capable of turning into a form with a higher free energy value, e.g., due to rearrangement or formation of new chemical bonds. This symbol can also designate a membrane capable of undergoing a transition into a new constrained state (e.g., due to a conformational transition or to a rise of a difference in the concentration of transmembrane hydrogen ions).

As was already stated in Sect.8.1 there exist two types of processes of ATP synthesis: substrate and membrane phosphorylation. The first process is carried out by isolated, mainly glycolitic, enzymes, and membrane phosphorylation takes place within organized membrane multi-enzyme structures (ETC) of mitochondria (oxidative phosphorylation) and chloroplasts (photosynthetic phosphorylation). In this section we shall consider substrate phosphorylation.

As an example let us discuss here the oxidation of glyceraldehyde-3-phosphate with coenzyme NAD^+ catalyzed by the glyceraldehyde-3-phosphate dehydrogenase (this enzyme will be designated as E_1). In the presence of another enzyme, phosphoglycerokinase (E_2), ATP synthesis takes place coupled with the first process. According to (8.16) the overall process can be written down as follows:

$$\begin{array}{c} H \diagdown \quad \diagup O \\ C \\ | \\ H - C - OH + NAD^+ + H_2O \\ | \\ CH_2OPO_3H \\ \text{glyceraldehyd-3 phosphate} \end{array} \qquad E_1 \quad E_2 \qquad ATP + H_2O$$

$$\begin{array}{c} HO \diagdown \quad \diagup O \\ C \\ | \\ H - C - OH + NADH + H^+ \\ | \\ CH_2OPO_3H \\ \text{3-phosphoglyceric acid} \end{array} \qquad \qquad ADP + P_i \tag{8.18}$$

This complex summary process passes a series of stages. In order to understand the meaning of physical problems arising in the course of its analysis, it is necessary to consider these stages in greater detail. A minute description of experimental data and the references can be found in an excellent monography [8.2].

Coenzyme NAD^+ forms with E_1 a strong complex (probably through the S atom of one of the protein sulphohydril groups)

$$E_1SH + NAD^+ \rightarrow E_1 - S - NAD + H^+ \quad . \tag{8.19}$$

The substrate (glyceraldehyde-3-phosphate) interacts with this complex, reduces NAD^+ and substitutes NADH forming the so-called acyl-enzyme

$$\begin{array}{c} H \diagdown \quad O \\ \diagup \\ C \\ | \\ H - C - OH \\ | \\ CH_2OPO_3H_2 \end{array} + E_1 - S - NAD \rightarrow E_1 - S \sim \overset{O}{\overset{\|}{C}} - \overset{H}{\overset{|}{C}} - CH_2OPO_3H_2 + NADH \quad . \tag{8.20}$$

In the acyl-enzyme, the substrate is already oxidized. In water medium, acyl-enzyme is subjected to slow hydrolysis, forming one of the end products of (8.18), 3-phosphoglyceric acid, and the enzyme is regenerated

$$E_1 - S \sim \overset{O}{\overset{\|}{C}} - \overset{H}{\overset{|}{\underset{OH}{C}}} - CH_2OPO_3H_2 + H_2O \rightarrow E_1 - S - H + HO - \overset{O}{\overset{\|}{C}} - \overset{H}{\overset{|}{\underset{OH}{C}}} - CH_2OPO_3H_2 \quad . \tag{8.21}$$

In this case, the energy liberated in the course of substrate oxidation by co-enzyme dissipates into heat. The acyl-enzyme hydrolysis proceeds, however, extremely slowly, and, therefore, at catalytic quantities of enzyme the glyceraldehyde-3-phosphate oxidation does not, practically, take place. In order to carry out (8.20) quantitatively in such artificial conditions, large concentrations of enzyme E_1 are required. E_1 in this case plays the role of a reagent and not of a catalyst. In the presence of inorganic phosphate, the enzyme turnover number increases more than 150000 times. Reaction (8.20) is, in this case, accompanied by process

$$E_1 - S \sim \overset{O}{\overset{\|}{C}} - \overset{H}{\overset{|}{\underset{OH}{C}}} - CH_2OPO_3H_2 + H_3PO_4 \rightarrow E_1 - S - H + H_2O_3P - O \sim \overset{O}{\overset{\|}{C}} - \overset{H}{\overset{|}{\underset{OH}{C}}} - CH_2OPO_3H_2 \tag{8.22}$$

as a result of which the enzyme is regenerated and the 1,3-diphosphoglycerate with a high-energy bond in position 1 is formed. The next, concluding stage of the process is the transfer of high-energy bond onto the ADP and the formation of 3-phos-phoglyceric acid (the catalyst is phosphoglycerate kinase, F_2)

$$H_2O_3P - O \sim \overset{O}{\overset{\|}{C}} - \overset{H}{\overset{|}{\underset{OH}{C}}} - CH_2OPO_3H_2 + ADP \overset{E_2}{\rightarrow} HO - \overset{O}{\overset{\|}{C}} - \overset{H}{\overset{|}{\underset{OH}{C}}} - CH_2OPO_3H_2 + ATP \quad . \tag{8.23}$$

The algebraic sum of (8.19,20,22,23) corresponds, naturally, to process (8.18).

As a matter of fact, the high-energy phospho-ether bond has already appeared after the formation of kinetically stable (in the absence of E_2) 1,3-diphospho-glycerate. The last stage (8.23) is the reaction of transfer of this bond onto ADP. We shall restrict ourselves to the consideration of the overall process up to and including stage (8.22)[6]. It has already been stated, that, in the absence of inorganic phosphate, enzyme E_1 behaves as a reagent and not as a catalyst. The complex of E_1 with NAD interacts with the substrate producing acyl-enzyme and NADH [reaction (8.20)]. The slow hydrolysis of acyl-enzyme is, in fact, a reaction decoupling the processes of glyceraldehyde-3-phosphate oxidation and phosphorylation. It is, therefore, clear, that acyl-enzyme is already a high-energy compound. E_1 acylation in water medium is energetically unfavorable, and acyl-enzyme hydrolysis leads to a decrease of the system free energy, the excess of which is being dissipated. It means that $S \sim \overset{\overset{O}{\|}}{C}$ - bond formation, in the course of (8.20) is, in the absence of phosphate, unfavorable and can be realized only because another, energetically favorable process of NAD oxidation proceeds *simultaneously*. The word "simultaneously" has been underlined deliberately. We shall try to specify below the meaning of "simultaneity" in this case.

The detailed mechanism of substrate phosphorylation during the interaction of acyl-enzyme with inorganic phosphate (8.22) is not completely clear. Phosphate is also known (see [8.2]) to form a complex with E_1. E_1 is a "two-headed" enzyme with two active centers, one of which is responsible for phosphate transfer, and the other for acylation. The interaction of substrate with enzyme and phosphate is, probably, realized in the course of one elementary act, and, in the presence of phosphate, acyl-enzyme does not exist as a kinetically independent unit. Whatever the concrete mechanism would be, it is clear that the formation of a phospho-ether band in position 1 [see (8.22)] is energically unfavorable in water medium and can take place either at the expense of a simultaneously proceeding thermodynamically favorable NAD reduction or as a result of phosphorylation of the already formed kinetically stabilized macroerg (acyl-enzyme). In this case, the word "simultaneously" means that the time interval between two elementary steps is so small that the energy liberated during the exoergonic process does not have enough time to dissipate into heat. Here, both processes are in fact taking place in the course of one elementary act (or, speaking in terms of the absolute reaction rate theory, have one and the same activated complex). In the case of conventional reactions of low-molecular compounds, the characteristic times are those of vibrational and rotational relaxations, i.e., $10^{-11} - 10^{-13}$ s. If this estimation is correct it is extremely improbable that there is taking place a complex process with the participation of three low-molecular compounds and at least two groups in two active centers of the protein macromolecule within one elementary act.

---

[6]It does not mean that the physical mechanisms of the change of one high-energy bond for another in (8.23) are completely clear. For them to be understood, however, no new problems arise, which could not be analyzed during the discussion of the preceding stages.

The mechanism of energy coupling of chemical reactions is the central problem of bioenergetics. A possible way for its solution is considered below.

In Chap.6 dealing with the physical aspects of enzymatic catalysis a new concept of the elementary act in enzymatic reactions was presented. According to this concept, addition of substrate to an enzyme active center leads to the apperance of a new kinetically available conformational state of the macromolecule.

The initial conformation becomes constrained and slowly relaxes to the new state. During this relaxation the catalyzed chemical transformation takes place. This slow relaxation is in fact mechanical motion. If the chemical transformation of the substrate is accompanied by considerable energy decrease (as, e.g., in the course of glyceraldehyde-3-phosphate oxidation and NAD reduction), we can say that during this slow conformational relaxation energy is being liberated. After substrate addition to the enzyme the unchanged, but now mechanically constrained, initial conformation of the macromolecular complex becomes the primary macroerg. The time interval, during which this energy can be utilized to perform an exoergonic process, is comparable to the relaxation times for specific mechanical degrees of freedom excited by substrate attachment ($10^{-5} - 10^{-1}$ s).

If, during conformational relaxation of the whole macromolecular complex, there is a change in electron density distribution in the active center responsible for the endoergonic process, both the exoergonic and the endoergonic, processes constitute one elementary act, although there may be an interval of tens and hundreds of milliseconds between them. Energy transfer from one process to the other takes place due to the excitation of slowly relaxing mechanical degrees of freedom of the macromolecular complex. The details of a possible mechanism of this process will be discussed in Sect.8.6 with membrane phosphorylation being used as an example.

8.4 Membrane Phosphorylation: Thermodynamic Aspects

The most important examples of membrane phosphorylation are oxidative phosphorylation in the inner membranes of mitochondria and photosynthetic phosphorylation (photophosphorylation) in the membranes of chloroplasts of higher plants and algae and in the chromatophores of photosynthetic bacteria. We shall begin with discussing the main experimental facts concerning oxidative phosphorylation in mitochondria. As usual, we restrict ourselves to the data which will be necessary to understand physical problems of general interest. A detailed description of mitochondrial structure can be found in [8.2,3,25].

The discovery of oxidative phosphorylation must be probably dated from 1930 when ENGELHARDT [8.26,27] established a connection between respiration and phosphorus exchange within erythrocytes. KALCKAR [8.28,29] got similar results with cell-free preparations, and BELITZER and CYBAKOVA [8.30] were the first to find the stoechiometric relationship between respiration and phosphorylation. These most important studies were continued by OCHOA [8.31,32] with cell-free preparations. The main result of the cited works was that they established the fact that the oxidation of one substrate molecule (i.e., the formation of one water molecule in reaction $2H+1/2O_2 \rightarrow H_2O$) is accompanied by the formation of more than one ATP molecules. We know now that, if the substrate of ETC (see Sect.7.3) is NADH, three ATP molecules are formed in undamaged "coupled" mitochondria per one atom of oxygen absorbed (i.e., per two electrons that have passed through the ETC). In other words, the ratio P/O = 3. If succinate is used as a substrate, P/O = 2, and for substrate phosphorylation the corresponding ratio equals unity. It means that the passage of electrons along the ETC is accompanied by ADP phosphorylation at several sites of ETC (so-called coupling sites).

ETC enzymes responsible for electron transfer and the phosphorylating enzymes necessary for ATP synthesis from ADP and phosphate are united in mitochondrial membranes. They are included into a complex lipoprotein structure and can only be isolated with great difficulty.

The overall process, being realized in coupled ETC in the course of NADH oxidation, can be written out as follows:

$$NADH + 3P_i + 3ADP + \tfrac{1}{2}O_2 + H^+ \rightarrow NAD^+ + 3ATP + H_2O \quad . \tag{8.24}$$

One can say with a rather high degree of certainty that ADP phosphorylation is realized by a specific enzyme-ATP-ase, which is not a part of ETC but is localized within the membrane near the ETC polyenzyme system. There are many works devoted to studying the mitochondrial ATP-ase and the proteins necessary for its connection with ETC components (so-called, coupling factors). Most chemical problems concerning the mitochondrial ATP-ase remain unsolved up to now. We do not even know if there exist different ATP-ases (one for every coupling site) or whether one and the same enzyme attends to all the coupling sites. The last assumption seems to be more verisimilar (see [8.2]).

One can distinguish five states of mitochondrial ETC [8.33-36]. State 1 is realized in the absence of externally added substrates of respiration and phosphorylation. The ETC carriers are partly reduced. The addition of ADP (substrate of phosphorylation) converts mitochondria into state 2: all carriers become oxidized, the respiration rate is determined by the substrate of oxidation. If we now add the substrate of oxidation, mitochondria are converted into active state 3, the carriers again become partly reduced, and the rate of electron transport is determined by the rate of substrates entering through the external membrane of mitochondria and by the activity of phosphorylating enzymes. When a greater part of added ADP has been spent (transformed into ATP), or when a great excess of ATP has been specially added, the state of respiration control (state 4) arises, respiration again slows down, and its rate is determined by the ADP/ATP ratio.

Changing this ratio, e.g., by ADP addition, one can again convert mitochondria from state 4 into the active state 3. In the absence of O_2, mitochondria are in state 5, i.e., in the state of anaerobiosis, and all ETC carriers become reduced.

This classification suggested by CHANCE turned out to be very useful. With the help of data concerning the influence of state 4 and state 3 transition upon the degree of oxidation of certain carriers, as well as those on the action of electron transport inhibitors and phosphorylation decouplers, CHANCE suggested a scheme for the localization of coupling sites in ETC. Up to now this scheme has not been modified to any essential. The question of coupling sites will be discussed later.

All the aforesaid leads to the conclusion that the main biophysical problem of the oxidative phosphorylation process is the formation mechanism of primary macroerg which produces ATP with the help of ATP-ase. Formulated in this way, the problem differs from that of substrate phosphorylation only by the nature of chemical reaction, which is accompanied by energy decrease and leads to the formation of primary macroerg. In the case of oxidative phosphorylation, this reaction is the transfer of an electron between fixed neighbor components of ETC.

Let us first of all discuss the thermodynamic aspect of this problem. It is clear that overall process (8.24) taking place in coupled mitochondria can be subdivided into two stages. Exoergonic stage

$$NADH + H^+ + \tfrac{1}{2}O_2 \rightleftharpoons NAD^+ + H_2O \tag{8.25}$$

is realized as a result of the transfer of two electrons through ETC from NADH to molecular oxygen. We can assume that initial and end products of stage (8.25) are equilibratedly distributed between the ETC and the medium. We are, therefore, justified in using the formulae of equilibrium thermodynamics when discussing the overall process. The free energy change in the course of these stages is equal to the difference between the redox potential values of $NADH/NAD^+$ and $1/2O_2/H_2O$ pairs. Within the cells, $NADH/NAD^+ \approx 1$, $pO_2 \approx 0.2$ atm, pH ≈ 7, and this difference is [8.37]

$$\Delta F \approx -51 \ kcal/mol \quad . \tag{8.26}$$

This energy can be used to perform various endoergonic processes, phosphorylation included. The cell can utilize $51/3 = 17$ kcal per one mole of ATP synthesized. MURAOKA and SLATER [8.38] have shown that in the state of respiratory control (state 4 according to CHANCE's classification) a thermodynamic equilibrium between the overall redox process and phosphorylation sets in after a certain value of $[ATP]/[ADP][P_i]$ has been reached within mitochondria. By measuring the concentrations of ATP, ADP and P_i in the medium for state 4, we can, therefore, calculate the free energy necessary to synthesize one mole of ATP

$$\Delta F' = \Delta F^0_{obs} + 1.36 \ \log \frac{[ATP]}{[ADP][P_i]} \quad , \tag{8.27}$$

where ΔF^0_{obs} is the standard free energy of ADP phosphorylation. At 25 C and physiological pH values, $\Delta F^0_{obs} \approx 8.0$ kcal/mol. Using data on ATP, ADP and P_i concentrations in mitochondrial medium for state 4 [8.39,40] we obtain

$$\Delta F' \approx 8.0 + 6.5 = 14.5 \text{ kcal/mol} \quad . \tag{8.28}$$

Thus, out of the available 17 kcal about 14.5 kcal are used for the synthesis of one mole of ATP, and the overall reaction of oxidative phosphorylation (8.24) in the state of respiratory control practically proceeds reversibly.

We must now recall (see Chap.7) that electron transfer between all ETC components is, probably, realized according to the one-electron mechanism. In order for the decrease in energy of electrons passing through the coupling site to be used for the synthesis of ATP molecules, the drop of redox potential at this site must be either not less than 630 mV (if one act of electron transfer leads to the synthesis of one ATP molecule) or not less than 315 mV (if two acts of electron transfer are necessary for the synthesis of one ATP molecule)[7]. In the last case, there must exist a mechanism of the accumulation of energy liberated in the course of two consecutive electron transfer acts.

Figure 8.1 shows the localization of coupling sites in mitochondrial ETC [8.41]. Coupling site I is localized between the group of carriers associated with NADH dehydrogenating flavoprotein and the ubiquinone UQ. Site II is localized near cytochrome b_T (see Sect.7.3), and site III is associated with the complex formed by cytochromes a_3, a and copper ions (see also [8.42]).

The difference between equilibrium values of redox potentials of any pair of neighboring carriers in ETC is smaller than 630 mV and even than 313 mV (see, e.g., [8.43]). This "energy paradox" is, probably, one of the central problems of membrane phosphorylation.

First of all, it must be clarified whether the equilibrium values of redox potentials (ROP) of isolated carriers can be used in the analysis of ETC energy properties. It would be quite reasonable to assume that ROP of ETC carriers fixed in a membrane may differ considerably from those of isolated carriers. Moreover, it is possible that ROP of carriers-transformers[8] may be essentially dependent on the membrane state, in particular, may undergo changes with membrane energization. On the other hand, it is not excluded that in such well-organized structures as carriers containing membrane ETC the energy liberated in the course of electron transfer differs considerably from the ROP difference of neighboring carriers. Confor-

---

[7]The transition from the primary macroerg to ATP is supposed not to be accompanied by energy dissipation. Otherwise, the required redox potential differences would be even greater.

[8]I.e., the carriers, whose redox transformations lead directly to the primary macroerg appearance.

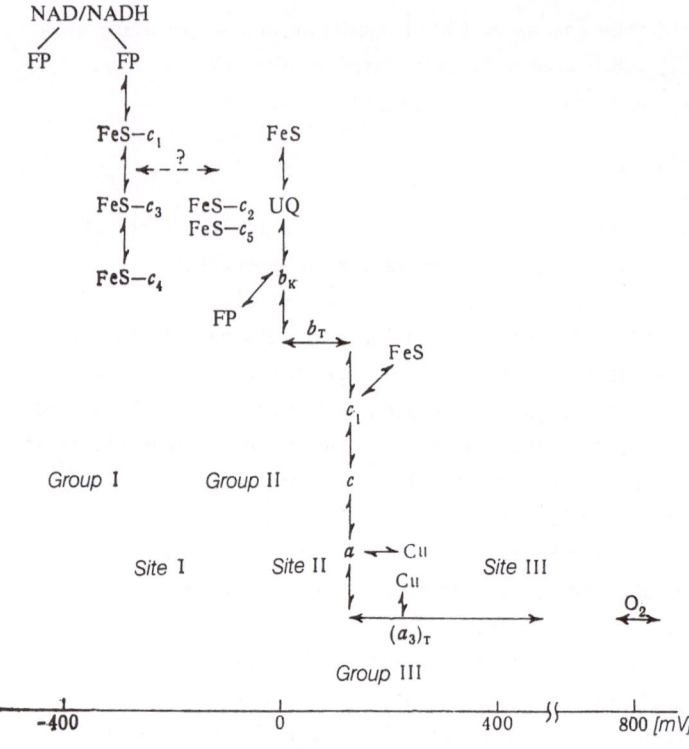

Fig.8.1. Localization of coupling sites in ETC of mitochondria. Index "T" indicates carriers which are energy transformers

mational relaxation of carriers is very slow, and the major part of the time during which they are functioning they are, therefore, in a kinetically out-of-equilibrium state. At the same time, the ROP values of carriers within the membrane may remain almost unchanged as compared with those for isolated carriers. It does agree with the fact that electron absorption spectra of isolated carriers and those of carriers in a membrane differ rather slightly. In order to choose one of these two alternative solutions of the energy paradox it is necessary to consider the data concerning ROP values of ETC carriers within the mitochondrial membrane. These measurements were carried out by WILSON and DUTTON [8.44,45]. ROP values of the medium were established by means of a suitable choice of oxidizing and reducing agents in buffer quantities, and N,N,N^1,N^1-tetramethyl-p-phenylenediamine (TMPD), phenazinemetasulphate or phenazineetasulphate were used as mediators. Concentration ratios of the carrier (cytochromes) oxidized and reduced forms were measured spectrophotometrically. With the help of this method coupling sites II and III were studied.

It was found that two cytochrome b components (called b_k and b_T [8.46]) in coupling site II do not behave in the same way after the energization or decoupling of mitochondria. The kinetical differences have been already discussed in Sect.7.3.

ROP values for cytochromes b_k and b_T in uncoupled or nonenergized mitochondria were measured to be +20 and -35 mV, respectively. Energization by ATP addition does not change the ROP value of cytochrome b_k, but leads to the increase of cytochrome b_T ROP by about 275 mV (from -35 to +240 mV). The authors concluded that there exist two forms of cytochrome b: its low-potential and high-potential forms. In the course of mitochondria energization, the low-potential form turns into the high-potential form. As a matter of fact, the high-potential form of cytochrome b_T is, according to these ideas, the primary macroerg of coupling site II, and can be designated as $b_T^\sim$. The change of ROP value of cytochrome b_T during the energization determines the energy, which can be utilized for the primary chemical high-energy compound (and then ATP) synthesis. CHANCE [8.41] proposed a concrete cycle of cytochrome b_T transformations during which the carrier undergoes subsequent transitions between all the four possible forms (b_T^{3+}, b_T^{2+}, $b_T^{3+}\sim$, and $b_T^{2+}\sim$). The numerical data given above indicate that ATP synthesis requires a parallel or subsequent transfer of two electrons through coupling site II[9].

These ideas have been developed in [8.47-49]. Similar methods were applied to study the dependence of ROP values of cytochromes a_3 and a on the energization state of mitochondria. It was found that after the energization of mitochondria by ATP addition the ROP value of cytochrome a_3 decreased from 395 to 300-290 mV. Decouplers cut off this effect. The standard ROP value of cytochrome a (+190 mV) does not change under ATP action. Using these data, CHANCE and ERECINSKA [8.50] suggested a scheme of energy transduction and the formation of primary chemical macroerg at coupling site III. The primary macroerg in this case is the low-potential form of cytochrome a_3 with ROP ~+300 mV. This interpretation is rather improbable, the changes of ROP in the course of energization being too small (the change of system free energy due to redox process at site III is about six times less than that required for the synthesis of one ATP molecule).

This interpretation of the results obtained by WILSON and DUTTON was criticized by CASWELL [8.51]. He stated that the redox mediator, TMPD, used by WILSON and DUTTON easily interacts in the mitochondrial membrane only with cytochrome c, very poorly with cytochrome b_1, and does not interact at all with cytochrome a and a_3 (see also [8.52]). Let us assume that the mediator ensuring the connection between redox processes in the medium and redox reactions of mitochondrial ETC carriers interacts only with cytochrome c, and that ETC intrinsic electron transport is negligible (WILSON and DUTTON used inhibitors). Here is the scheme of the corresponding ETC region

$$b_T \rightleftharpoons c_1 \rightleftharpoons c \rightleftharpoons a \rightleftharpoons a_3 \quad .$$
$$\uparrow\downarrow$$
$$M \text{ (mediator)}$$

---

[9]Strictly speaking, even in this case the energy obtained (~0.55 eV) would not be enough to ensure the synthesis of one ATP molecule (~0.63 eV).

If there is at least one coupling site between cytochrome c and cytochrome b, as well as between c and a_3, it follows that the concentration ratios of their oxidized and reduced forms at given ROP of the medium are not simple functions of their standard ROP values. Let us assume that an equilibrium has been established between electron transfer through a coupling site and ATP synthesis (at stoichiometric proportion: 2 electrons per 1 ATP molecule). In this case, the following equations should be valid for coupling sites II and III

$$K_{II} = \frac{[b_T^{ox}]^2[TMPD^{red}]^2[ATP]}{[b_T^{red}]^2[TMPD^{ox}]^2[ADP][P_i]} \tag{8.29}$$

$$K_{III} = \frac{[a_3^{red}]^2[TMPD^{ox}][ATP]}{[a_3^{ox}]^2[TMPD^{red}][ADP][P_i]} . \tag{8.30}$$

The measured $[a_3^{ox}]/[a_3^{red}]$ and $[b_r^{ox}]/[b_T^{red}]$ ratios for coupled mitochondria will change with ATP addition in accordance with (9.29,30), and not due to the changes of state and the changes of standard ROP values of carriers. Indeed, CASWELL has shown that, with mediators directly interacting with cytochrome b (vitamin K and 2,6-dichlorophenolindophenol), the standard ROP value for this carrier in mitochondria is equal to +75 mV and does not change under the action of ATP or uncouplers. According to these results, no specific high-potential form of cytochrome b exists which is at equilibrium for the given state of mitochondrial membrane. In the case of a mediator directly reacting with a carrier, the measured ratio between the concentrations of the oxidized and reduced carrier forms is equal to the ratio between the oxidation and reduction rates for this carrier and corresponds to the true redox equilibrium. On the other hand, for a mediator reacting directly with another carrier separated from the carrier in question by a coupling site, the oxidation or reduction rate will be determined by the slow process of ATP synthesis. The measured concentration ratio, although being equal to the ratio between the reduction and oxidation rates, will not, therefore, reflect the true redox equilibrium of this carrier.

It is appropriate to make here one more remark. The ratio between the concentrations of oxidized and reduced carrier forms was measured in these experiments spectrophotometrically, i.e., mainly determined by the redox state of prosthetic group. In Chap.5 we have, however, seen that oxidation and reduction of cytochromes can be accompanied by considerable conformational changes, proceeding, as a rule, more slowly than the changes of the prosthetic group. The measured concentration ratios may not, therefore, correspond to the concentration ratio of equilibrium relaxed oxidized and reduced forms. We shall see later (Sect.8.6) that, with this method of measurement, the rate of conformational changes may become the factor determining the change in the carrier redox state, due to membrane energization (and, consequently, the measured change in the system free energy). The measured values

will strictly correspond to the energy liberated during electron transport and
available for ATP synthesis only in the case of an infinitely slow transformation.
In all other cases, one would obtain underestimated values. As a matter of fact,
conventionally measured standard ROP values of ETC carriers are either not standard
ROP in the true sense of the word (and their changes due to energization are always
smaller than the true changes in the system energy), or they are equilibrium values
(if measured by Caswell method), but their difference for neighboring carriers does
not correspond to the true energy changes which accompany the electron transfer in
a kinetically out-of-equilibrium coupled membrane.

In recent years the number of papers concerning photophosphorylation in photosynthetic systems has been growing incessantly, but, nevertheless, the coupling sites
in chloroplasts and chromatophores have not been established precisely. One may
regard as proved the fact that photophosphorylation is a dark process coupled with
electron transfer between fixed carriers in ETC of chloroplasts or chromatophores
(see Sect.7.3). However, JOLIOT et al. [8.53] measured the equilibrium constant of
dark reversible electron transfer between active centers of photosystems I and II
in chloroplasts of green plants and chlorella and found that this constant does not
depend on the presence of ATP and decouplers. These authors concluded that electron
transfer between the two photosystems is not coupled with phosphorylation, which
is, probably, energetically ensured directly by the light quanta absorption. We have
already seen, however, that the so-called "equilibrium thermodynamic properties of
centers and carriers fixed in membranes" usually do not have any relation to the
actual process. The data concerning the number and the characteristics of a coupling site in noncyclic ETC can be found in [8.54]. AVRON and CHANCE [8.55] have con-
vincingly shown that at least one coupling site is localized in ETC between two
photosystems before cytochrome f. Interesting results were published in 1970 by
HAUSKA et al. [8.56]. They obtained subchloroplast particles capable of carrying
out the photoreduction of NADP and cyclic photophosphorylation. These particles
(mainly containing photosystem I) can perform cyclic photophosphorylation only in
the presence of an exogenous reducing agent which supplies a certain amount of
electrons from the outside. It is possible that, in the initial chloroplasts, the
role of such an exogenous donor for cyclic photophosphorylation is played by photosystem II. It was indeed shown quite long ago [8.57] that chlorophenyldimethyl-
urea — an inhibitor of inter-system electron transport — inhibits the cyclic photophosphorylation as well. In any case, the results of [8.56] can be interpreted as
an indication of parallel transfer of two electrons being necessary for the primary
chemical macroerg formation (see later, Sect.8.5). The authors [8.56] concluded
that, in cyclic phosphorylation, the coupling site is localized between P 700 and
cytochrome b_6 [according to their scheme, the latter closes the cycle and directly
reduces $(P\ 700)^+$ bypassing cytochrome f and plastocyanine (PC)]. The coupling site
of noncyclic phosphorylation is localized between plastoquinone (PQ) and cytochrome

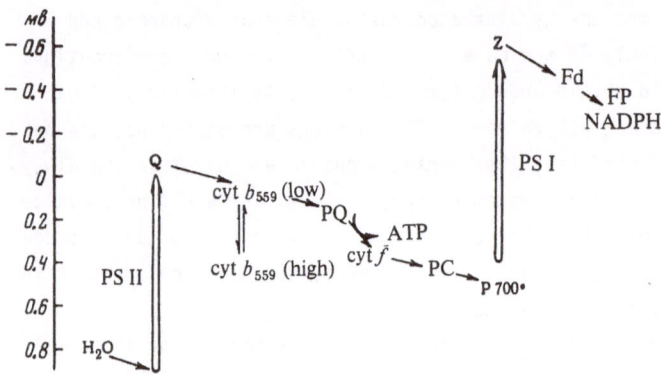

Fig.8.2. Localization of coupling site in noncyclic ETC of chloroplasts [8.58]

f. BOHME and CRAMER [8.58], who studied the kinetics of electron transport and phosphorylation in chloroplasts, reached the same conclusion. According to their data, cytochrome f and plastoquinone are neighbors in noncyclic ETC (Fig.8.2), and electron transfer between them is a limiting stage of the overall electron transport in coupled chloroplasts. The authors thought that it is the sole coupling site in the whole noncyclic electron transport path. The difference between the standard ROP values of plastoquinone and cytochrome f is about 250 mV, and the passage of one or even two electrons through this site cannot give enough energy to synthetize one ATP molecule. In the course of noncyclic phosphorylation, the number of ATP molecules synthetized per two electrons passing from water to NADP is close to two [8.59]. It means that either there exists at least one more coupling site, or the true decrease of electron energy, when it is transferred from PQ to cytochrome f, approaches 0.63 eV and does not correspond to the difference between the equilibrium ROP values.

In their study of electron transport and photophosphorylation in chloroplasts with various electron acceptors SAHA et al. [8.60] reached the conclusion that, in the normally functioning chloroplasts, the P:2e ratio is equal to two (see also [8.61]), and that there exist two separate coupling sites, one of which is, probably, localized between water and the active center of photosystem II. The same suggestion was made earlier in [8.62]. The drop of standard ROP values between the water and the photosystem II active center does not, probably, exceed 100 mV, and the thermodynamics of this process remains obscure in the realm of the classical approach.

Thermodynamic aspects of photophosphorylation in chloroplasts and oxidative phosphorylation in mitochondria are rather similar. The mechanisms of these two processes are, probably, also very much alike (see, e.g. [8.63]).

We have only discussed in this section the experimental data concerning the energetics of overall process of membrane phosphorylation. In Sect.7.3 the main result of kinetic studies, namely, the retardation of electron transport by the coupling process, was already considered rather comprehensively. Naturally, the studies of membrane phosphorylation carried out for many years in many laboratories in different countries have lead to discoveries of many other experimental facts and empirical rules, concerning, for instance, pH difference across the membrane, transmembrane potential, the action of various uncouplers and inhibitors, etc. It is impossible to present these data in detail here. We shall discuss some of them in the next section as related to certain theoretical concepts of membrane phosphorylation.

8.5 Membrane Phosphorylation: Existing Theories

All the presently existing points of view on the mechanism of membrane phosphorylation (more exactly, on the nature of the primary energy transduction process) can be subdivided into three main groups: chemical, chemiosmotic, and conformational concepts [8.37]. As we have already stated, their common feature is the assumption of the existence of a primary macroerg not identical to ATP. These concepts differ as to the notions of the nature of primary macroerg and the mechanism of the primary process of energy conservation.

In its present form, the chemical concept was formulated by SLATER in 1953 [8.64]. This concept is based on the assumption that the primary act of energy conservation, i.e., the process of the transformation of energy liberated, as a result of electron transfer, into a long-living form, takes place within the molecule of a "carrier-transformer", and is essentially the formation of a high-energy chemical compound, containing one of the redox reaction products and a low-molecular ligand (denoted below as "I")

$$AH_2 + B + I \rightleftharpoons A \sim I + BH_2 \quad . \tag{8.31}$$

After that, this primary macroerg ensures ATP synthesis with the help of ATP-ase enzyme

$$A \sim I + ADP + P_i \rightleftharpoons A + I + ATP \quad . \tag{8.32}$$

As a matter of fact, the chemical concept of membrane phosphorylation differs from the explanation of substrate phosphorylation only by the assumption that the redox process comes to the one-electron transfer between neighboring ETC components fixed within the membrane. The physical mechanism of (8.31) remains obscure. In order to prevent the total energy dissipation, electron transfer from A to B and the formation of high-energy bond $A \sim I$ must be realized in one elementary act. As was already stated in Sect.8.3, this is very difficult to conceive in the realm of classical notions. It is easy to see that the chemical concept of membrane phosphorylation is a natural extension of the approach to substrate phosphorylation presented in Sect.8.3.

The conformational concept is based on a multitude of experimental data concerning conformational changes of protein carriers in the course of redox processes and those in the whole membrane in the course of its functioning. According to this concept, chemical macroerg formation may be conceived as the reverse of myosin conformational change during ATP hydrolysis. The primary act of energy conservation is supposed to be the conformational change of the respiratory carrier (favorprotein, iron-sulphure protein or cytochrome) induced by electron transfer [8.65].

A detailed discussion of conformational hypothesis can be found in the articles by GREEN and co-workers [8.66,67]. The general scheme of ATP synthesis can be presented as follows:

$$\text{Electron transfer} \rightleftharpoons \text{A conformationaly} \rightleftharpoons \text{ATP}$$

changed energy-rich
state   (8.33)

The authors of [8.67] considered the energy-rich conformationally changed state of the whole membrane. It is assumed that a membrane can exist either in an energy-poor or in an energy-rich state. These states are spreading along the membrane (from one lipoprotein complex of the ETC to another) as a kind of a peculiar wave. It seems that, according to this concept, the primary act is the energization (an induced conformational change) of macromolecular carriers. From this point of view, the conformational concept may be regarded as a variant of the chemical concept, the primary chemical macroerg, $A \sim I$, being replaced by the energy-rich conformationally changed state of carrier A. At the same time, the assumption of an energy-rich state of the whole membrane, where the energy liberated in the functioning ETC is conserved, makes the conformational concept similar to the one chemiosmotic.

The first variant of the chemiosmotic concept was proposed by MITCHELL in 1961 [8.68]. His hypothesis was based on the assumption that, in certain regions of ETC, the electrons are transferred across the membrane leading to the rise of ΔpH between its inner and outer surfaces. Thus, substrate oxidation on the outer surface

$$SH_2 \rightarrow S + 2e + 2H^+ \tag{8.34}$$

leads to proton accumulation in the outer space. The oxidation of the end acceptor (molecular oxygen) with the rise of proton deficit takes place on the inner surface, where cytochrome-oxidase is localized. Thus, between the two membrane surfaces appeares a difference of potentials or the membrane potential. A membrane with separated charges in fact represents, according to this concept, a primary macroerg. Protons, adsorbed on the surfaces, can be substituted by other cations with the surface charge being preserved and the outer space acidity increased. Electron transfer must, therefore, lead to two effects: the appearance of membrane potential and the difference of pH values at both sides of the membrane. The membrane potential can lead directly to ATP synthesis, for instance, according to the mechanism reverse to that of ATP-dependent active transmembrane transport of ions (incidentally, the physical foundations of this mechanism are rather obscure). Indeed, the reversal of Na^+ and K^+ motion across the membrane induces Na, K-dependent ATP-ase to catalize ATP synthesis [8.69].

In MITCHELL's initial scheme [8.68] phosphorylation was realized directly under the action of pH gradient resulting from electron transport. Later MITCHELL modified his scheme rather radically, and included certain intermediate stages in it [8.70]. The principles of chemiosmotic coupling can be seen in Fig.8.3. ETC functioning leads to the transfer of protons across the membrane, caused, in the end, by the difference of ROP values of ETC carriers. There exist the "coupling intermediates" XH and YOH of unknown nature within the membrane. XH and YOH are capable of being ionized and of undergoing a condensation reaction

$$XH \rightleftharpoons X^- + H^+ \tag{8.35}$$

$$YOH \rightleftharpoons YO^- + H^+ \tag{8.36}$$

$$YOH + XH \rightleftharpoons X - Y + H_2O \ . \tag{8.37}$$

The increase of the electrochemical potential of protons on the outer side of the membrane, due to the increase of their concentration and of the membrane potential, results in a shift to the right of the equilibrium

$$2H^+ + X^- + YO^- \rightleftharpoons H_2O + X - Y \ . \tag{8.38}$$

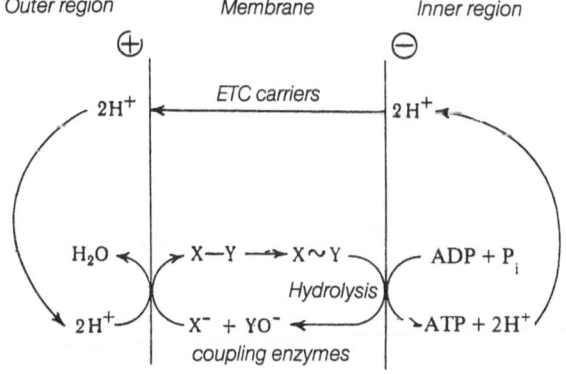

Fig.8.3.
Scheme of chemiosmotic coupling

The neutral X-Y molecule migrates across the membrane (not necessarily one and the same molecule passes through the whole membrane thickness). Contacting with the inner membrane surface, where the electrochemical potential of protons and the

concentrations of X⁻ and YO⁻ ions are low, X-Y becomes labile and the bond between
X and Y-energy-high (X~Y). The equilibrium (8.38) shifts to the left, ATP is synthetized at the expense of X~Y hydrolysis, and the liberated protons close the
cycle. Three types of coupling enzymes are necessary: a hydrolase, catalizing the
process on the outer membrane surface; translocase, ensuring transmembrane transfer; and synthetase, catalizing ATP synthesis. In order to satisfy the experimental
data on the existence of three coupling sites, it is necessary to assume the existence of at least three loops within ETC, where electron transfer is realized across
the membrane from the outside to the inside, and the proton transfer in the opposite
direction (see, e.g., Fig.8.4). The chemiosmotic concept has initially been developed as a natural reaction to the failure of attempts to find the primary chemical
high-energy compounds postulated by the chemical concept. The development of the
chemiosmotic concept has, however, made it necessary to postulate the formation of
coupling intermediate of unknown nature within the membrane which also precedes ATP
synthesis. The only difference between the chemiosmotic concept on the one hand and
the chemical and conformational concepts on the other is, thus, the assumption that
electron transfer through any coupling site leads directly to the appearance of a
difference in proton electrochemical potentials on both sides of the membrane
(ΔpH + membrane potential) and that the primary macroerg is, therefore, the changed
state of the membrane itself. At the same time, according to other concepts, the
primary macroerg is the changed (chemically or conformationally) state of a carrier[10].

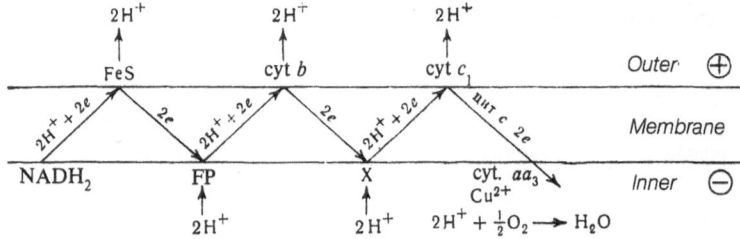

Fig.8.4. A possible scheme of electron and proton flow in mitochondrial ETC
according to chemiosmotic hypothesis

   Let us now discuss the arguments for and against these three concepts in
greater detail. The main shortcoming of these concepts is, probably, the fact that
neither of them answers the principal question concerning the physical mechanisms
of coupling. The chemical hypothesis reduces the problem of membrane phosphorylation to the no less obscure problem of substrate phosphorylation. The failure of

[10]A model unifying the conformational and chemiosmotic concepts has been recently
suggested [8.71].

attempts to isolate the primary chemical high-energy compounds must not embarrass anybody. It is quite possible that these primary macroergs are present within membranes as intermediates in negligible concentrations. Much more important is the fact that the measured ROP of intermediate carriers are not large enough to ensure ATP synthesis (see Sect.8.4). SLATER [8.37] proposed a coupling scheme based on the participation of a complex of two one-electron carriers acting as an energy transformer. In coupling site II in mitochondrial ETC the complex of two cytochromes b and b_i[11] may serve as this transformer. These cytochromes may exist in high-energy forms which differ from the conventional forms by their spectral and kinetic properties. SLATER's cycle for coupling site II in the mitochondrial membrane is shown in Fig.8.5. Only the oxidized cytochrome b_i and the reduced cytochrome b may exist in high-energy forms. Cytochrome b_i is reduceable in the high-energy state only. In the initial state of a dimer complex both these cytochromes are oxidized (b^{3+}, b_i^{3+}). The cycle includes seven stages. During stage 1 cytochrome b^{3+} is reduced by the electron transferred from the preceding ETC carrier (donor) and transformed into high-energy cytochrome $b^{2+}\sim$. In the course of stage 2 the high-energy state is transferred to the second component of the complex, b_i^{3+} which can now be reduced by the donor. The reduction takes place during stage 3, and the high-energy state is again transferred to cytochrome b^{2+}. In the course of stage 4 cytochrome b_i^{2+} is oxidized and energized transferring one electron to the next ETC carrier, i.e., to the acceptor. Both cytochromes are now in high-energy states ($b^{2+}\sim$, $b_i^{3+}\sim$). The next stage 5, is intracomplex electron transfer (the ROP of $b_i\sim$ is higher than that of b $\sim$) accompanied by de-energization of both cytochromes, the liberated energy being used for ATP synthesis. Stage 6 is intracomplex electron transfer in the backward direction (ROP of b_i is lower than that of b). The last stage, stage 7, is cytochrome b^{2+} oxidation and closes the cycle. In this ingeious scheme, however, the most important details remain obscure. These are: the nature of the high-energy state, the mechanism of its formation, and the mechanism of energy transfer to the ATP system. SLATER supposed that similar cycles can be formed for other coupling sites of the mitochondrial membrane (Fig.8.6).

The conformational concept is based on experimental data concerning the structural changes in membranes of coupled chloroplasts and mitochondria after electron transfer initiation or energization by ATP addition. These data were obtained with the help of various physical methods and they indicate the macrostructural, as well as local, changes in the membranes. Details and references can be found in [8.72-78]. There is, probably, no doubt that energy-transducing membranes and

---

[11]Cytochrome b_i in SLATER's notations is equivalent to cytochrome b_T in CHANCE's [8.41,46] notations (see Sects.7.3 and 8.4) as to their spectral properties. If, however, we compare their kinetic characteristics, the high-potential form of cytochrome b_T is equivalent to $b^{2+}\sim$ [8.37].

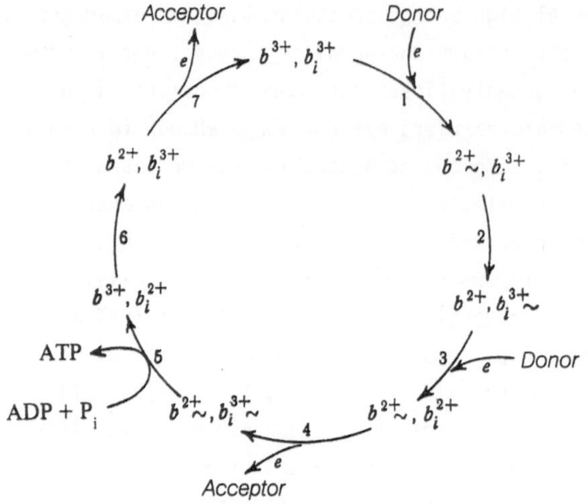

Acceptor   Donor

b^{3+}, b_i^{3+}

7   1

b^{2+}, b_i^{3+} $b^{2+}_{\sim}, b_i^{3+}$

6   2

b^{3+}, b_i^{2+} $b^{2+}, b_{i\sim}^{3+}$

ATP   5   3   Donor

ADP + P_i $b^{2+}_{\sim}, b_{i\sim}^{3+}$ 4 $b^{2+}_{\sim}, b_i^{2+}$

Acceptor

Fig.8.5.
Scheme of coupling site II in mitochondrial ETC

$$\text{NADH} \nearrow \text{FP} \rightarrow \text{FeS} \rightarrow \text{UQ} \rightarrow b \rightarrow c_1 \rightarrow c \rightarrow \text{Cu}, a_3 \rightarrow \tfrac{1}{2}O_2$$

$$\searrow \text{FeS} \rightarrow \text{UQ} \rightarrow b_i \longrightarrow c \rightarrow \text{Cu}, a$$

I   II   III

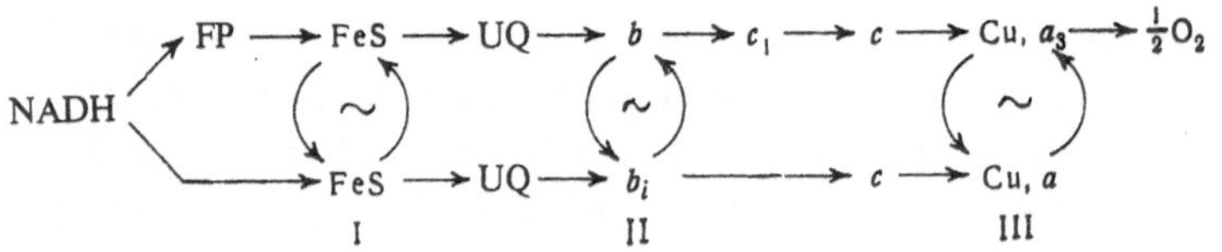

Fig.8.6. Coupling cycles in coupling sites I, II and III in mitochondrial ETC [8.37]

certain complexes within these membranes undergo structural transformations causatively associated with energy transduction. The authors of conformational hypotheses usually assume, however, that what is actually energized is the membrane structure, changed either directly as a result of electron transfer or under the action of electric field caused by electron transfer [8.66,67,79,80]. It is rather difficult to picture a mechanism of directional change of a structure, accompanied by an increase of its potential energy, as a result of a local electron transfer process. The physical mechanism of a chemical macroerg synthesis at the expense of the energy of a conformationally changed structure also remains unclear.

Most heated discussions in recent years have been provoked by Mitchells' chemiosmotic hypothesis. Three types of experimental data supporting this concept seem to exist. It is, first of all, the appearance of a pH gradient and transmembrane potential when membranes of mitochondria are energized by the substrate or ATP, and the coupled membranes of chloroplasts by light or ATP (see, e.g. [8.81-93]). Secondly, experiments by JAGENDORF and URIBE [8.94,95], who succeeded in performing ATP synthesis in chloroplasts by way of transition from acid to alkaline medium. Thirdly, it has been shown by many authors that the majority of uncouplers of membrane phosphorylation increase the membrane proton conductivity, thus leading to the decrease of ΔpH and membrane potential values (see, e.g.[8.37,96-99]).

The most attractive feature of the chemiosmotic concept at one time was the possibility to avoid intermediate chemical high-energy compounds which nobody could find. This advantage is now absent because MITCHELL was forced to assume the existence of an intermediate anhydride X-Y and the corresponding anions of unknown nature. The rise of ΔpH and membrane potential cannot in itself be regarded as proof of chemiosmotic hypothesis. Both these gradients can appear not as a cause but as a consequence of the processes leading to ATP formation. We have, indeed, seen in Chap.5, that the change in redox states of certain carriers may be accompanied by essential conformational changes in proteins which, in turn, lead, as a rule, to the changes in pK values of certain acidic groups, and consequently to the ejection or absorption of protons. If the distribution of these carriers is not symmetrical with respect to the outer and inner sides of the membrane, these changes must lead to the rise of the gradients observed. In this case, ΔpH and membrane potential may be considered only as characteristics of energized states of certain carriers of of the whole membrane. It is well known that many uncouplers of membrane phosphorylation simultaneously increase the membrane proton conductivity thus equalizing proton concentrations on both sides. This, however, cannot be regarded as proof of the suggestion that energization is reduced to just the creation of nonequilibrium distribution in which the increased potential energy of the system is stored. The kinetically out-of-equilibrium construction, such as individual ETC components or the whole membrane, possess, as a rule, "the triggers", which, having been pulled, can lead to a decrease of the system potential energy that had been localized on completely different degrees of freedom. Finally, direct evidence against chemiosmotic concept is, probably, presented by the results of measurements of membrane potential and ΔpH values at different states of coupling membranes. Nobody succeeded in obtaining the values of these parameters high enough to ensure ATP synthesis, even assuming dissipativeless energy transformation. Moreover, the membrane potential value did not, in certain cases, depend on the membrane energization state [8.100]. AVRON [8.101] gave a detailed analysis of this problem. He notes that the experiments by JAGENDORF and URIBE [8,94,95] are, indeed, evidence of it being in principle possible for the pH gradient to be used as the sole energy source for ATP synthesis in chloroplasts. It does not exclude, however, the fact that artificial creation of a huge pH gradient may lead to the apperance of a certain high-energy state ensuring ATP synthesis. Two experimental approaches can be applied to analyze the situation arising during photophosphorylation at normal physiological conditions. Firstly, one can measure the $H^+/h\nu$ or H^+/e^- ratio, i.e., the number of protons transferred across the membrane for each quantum absorbed, or for each electron transported along the ETC. According to the chemiosmotic concept, H^+/e^- must be equal to the number of coupling sites, i.e., to 1 or 2 for chloroplasts. Experimental results obtained by different authors disagree to a rather great extent. KARLISH and AVRON [8.102,103] have shown that, depending on conditions, one can obtain any value from 0.1 to 6.0. The quantum yield of proton transfer, $H^+/h\nu$, must be close to H^+/e^-. Direct measurements, carried out in two laboratories [8.104,105] gave, however, $H^+/h\nu$ values from 5 to 7, which do not agree with the chemiosmotic concept. Direct measurements of photoinduced membrane potential have given the values not exceeding 10 mV [8.91]. At the same time, it was shown that ATP synthesis can proceed at the zero value of membrane potential [8.106]. It was also shown that, at optimal conditions, the pH difference across the chloroplast membrane cannot exceed 3.5 [8.91]. This corresponds to about 200 mV and is not large enough to ensure ATP synthesis.

These conclusions based on experimental data cannot be regarded as irrefutable. Measurements of the true values of transmembrane potential in subcellular particles are very difficult to perform. If membrane potential values ar too low there are always doubts in the correctness of measurements. The small value of membrane capacitance leads to the creation of considerable potential differences without any great differences in surface charges and ion concentrations manifesting themselves on both sides of the membrane.

According to the present-day understanding, the chemiosmotic concept can be reduced to the following principal postulates. The transformation of energy, initially stored in the form of the differences between ROP of the carriers asymmetrically situated within the membrane, proceeds through a compulsory stage, as a result of which the energy is stored in the form of the transmembrane electrochemical potential of protons (μ_H+). This "charging" of the membrane condenser is the universal process of bioenergetical transformations in membrane structures [8.107]. Energy stored in the condenser can be utilized in the course of condenser discharge through a special construction-ATP-synthetase[12]. In the ATP-synthetase, protons give back their surplus energy (equal to the differences between their electrochemical potentials on both sides of the membrane), and ATP is synthetized. In order to get 1 ATP molecule at physiological conditions within the cell, 2-3 protons must pass through ATP-synthetase.

Thus formulated, the chemiosmotic hypothesis is almost irrefutable. It is rather difficult to establish whether the membrane potential is the cause or the consequence of some intermediate macroerg appearance. The only experiment directly contradicting the chemiosmotic concept would be ATP synthesis by means of ATP-synthetase in a homogeneous system. Nobody has done it as yet.

New experimental and theoretical works have called the principal propositions of the chemiosmotic concepts in question. Thus, NICHOLLS [8.108] simultaneously determined pH and transmembrane potential across the inner membrane of mitochondria, and concluded that their values do not satisfy the requirement of chemiosmotic hypothesis. YAMOMOTO and TONOMURA [8.109] reproduced the above-mentioned results

---

[12]We shall assign this name to the ATP-ase of mitochondria or chloroplasts.

[8.95] on ATP formation in the course of acid-base transitions in chloroplasts. It was shown that ATP synthesis is in this case determined not by transmembrane ΔpH values but by absolute proton concentrations. WITT and his co-workers proposed a new method of measuring the membrane potential, based upon the photoinduced changes in carotenoid spectra of chloroplast and mitochondria membranes (see, e.g. [8.110, 111]). With the help of this method numerous results were obtained, which speak in favor of the chemiosmotic hypothesis. The recorded spectral changes are considered to be caused by the electric field (electrochromism). It should be, however, noted that any weak perturbations must lead to similar spectral changes of the dye molecules. Irrespective of their cause, these changes can hardly allow us to make any conclusions concerning energy transformations. BALTSCHEFFSKY [8.136] showed that only a comparatively small part of these changes can be associated with photophosphorylation in bacterial chromatophores.

A critical review of the general tenets of chemiosmotic concepts can be found in [8.112-114]. Excellent reviews on experimental data of and on theoretical approaches to membrane phosphorylation were published recently [8.131-133].

8.6 Some Physical Aspects of Intracellular Energy Transformation as a Relaxation Process

In our discussion of the intracellular energy transformation, i.e., a process in the course of which the energy liberated during one chemical reaction is used to increase the potential energy of reagents of the second chemical reaction, we shall choose as an example the coupling between electron transfer and phosphorylation in mitochondria or chloroplast membranes. The principles upon which our approach is based were, probably, for the first time formulated in [8.115,116]. To begin with, let us forget the membrane as a whole and consider a coupling site consisting of three successive carriers A,B,C (electron transfer direction from A to C), each being a protein macromolecule with an active center capable of undergoing one-electron oxido reduction (e.g., a cytochrome).

Let us also assume at first that the energy difference (ΔE)[13] between the highest occupied electron levels of the reduced forms of A and C is large enough to provide for ATP synthesis ($\Delta E \geq 0.63$ eV). What is the meaning of this statement? We have already seen that any chemical (local) changes in a protein molecule especially in its active center lead to conformational transitions in the globule which, in turn, affect the state and the electronic characteristic of the active

---

[13]For the time being we shall not speak about ROP of carriers. We have seen that ROP measurements in functioning membranes are rather difficult to perform. Later we shall return to this question.

center. The change in the active center redox state causes changes in molecular conformations followed by additional changes in the active center electron level, whose position is, therefore, dependent on the degree of completion of the conformational relaxation. It is, therefore, necessary, when we discuss the energy difference between the levels of A and C carriers, to indicate what states we are speaking about. As we shall consider one-electron transfer only, a carrier can exist in two forms: the oxidized (e.g., A) and the reduced (e.g., A^-), each being either conformationally relaxed or not. The above mentioned energy difference, ΔE, relates to the relaxed reduced forms of carriers.

As we have seen, in order that an exoergic reaction should be able to energetically provide an endoergic chemical reaction, both these reactions must proceed in one elementary act. The elementary act of any chemical process involving a protein macromolecule is its conformational relaxation stage (Sect.6.5). Therefore, we have a right to assume, that the main energy changes in the carrier-transformer (carrier B in this case) are realized during this very stage.

Thus, the initial state of a three-carrier system: donor A, transformer B, and acceptor C, is A^-BC. Let us designate the relaxed (equilibrated) conformational state of the oxidized and reduced carriers with subindices "1" and "2", respectively, and the reduced states of their active centers with the superscript "minus". Let all the carriers be initially in conformational equilibrium

$$A_2^- B_1 C_1 \ . \tag{I}$$

Consider now the behavior of carrier-transformer in the course of electron transfer from A to C. This transfer proceeds according to the tunnel mechanism (Sect.7.5) and requires the coincidence of electron levels of the neighboring carriers within ~ 0.1 eV, this excess energy difference dissipating into heat by means of conventional vibrational relaxation. Therefore, the energy difference between the highest occupied electron level of the A_2^- active center and the lowest free electron level of the B_1 active center has to be not greater than ~ 0.1 eV. The transfer of an electron is followed by a fast vibrational relaxation of the now reduced active center and the close-lying protein groups, but the main part of the B macromolecule remains in the initial "oxidized" conformation, and the A^- macromolecule in the "reduced" conformation with an oxidized active center)

$$A_2 B_1^- C_1 \ . \tag{II}$$

Transformer B is now in conformationally unequilibrated (mechanically strained) state B_1^-. This state is strained not due to the change in the macromolecular conformation, but because electron addition to the active center of B has led to the appearance of the kinetically available conformational state B_2^- with a lower energy. Electron transfer from B_1^- to C_1 is forbidden because the energy difference between the corresponding electron levels is considerably greater than 0.1 eV. The slow conformational relaxation $B_1^- \to B_2^-$ must lead to the coincidence of the highest occupied level of B_2^- active center with the lowest free level of C_1 active center, the gap not exceeding 0.1 eV.

It is natural to assume that the $B_1^- \to B_2^-$ relaxation time is also large enough for the completion of the $A_2 \to A_1$ relaxation A and C are not transformers, and, therefore, their conformational relaxation must be accompanied by smaller energy changes.

The next state of the system may, thus, be written as

$$A_1 B_2^- C_1 \quad . \tag{III}$$

This state, in spite of great changes in the transformer conformation, is again conformationally equilibrated. The electron levels of B_2^- and C_1 now coincide with a high enough accuracy to allow the fast electron tunnelling with the formation of the state

$$A_1 B_2 C_1^- \quad . \tag{IV}$$

In this state, the conformations of B and C carriers are again unequilibrated and slowly relax to the final equilibrium state

$$A_1 B_1 C_2^- \quad . \tag{V}$$

The whole process can be written out as follows:

$$A_2^- B_1 C_1 \xrightarrow[a]{\text{fast}} A_2 B_1^- C_1 \xrightarrow[b]{\text{slow}} A_1 B_2^- C_1 \xrightarrow[c]{\text{fast}} A_1 B_2 C_1^- \xrightarrow[d]{\text{slow}} A_1 B_1 C_2^- \quad . \tag{8.39}$$

As a result, the electron is transferred from the donor A to the acceptor C, and the transformer B undergoes the cycle of oxido reduction and conformation changes.

The reduction and oxidation of the active center of transformer B take place during fast stages a and c. The initial states of B in these stages are conformationally equilibrated, and the comparatively small quantities of energy, released during these states, dissipate. Large energy changes, which can be, in principle, utilized to ensure some endoergonic chemical reaction, are released only during the slow relaxation stages b and d.

We now see that the discussion concerning the redox stage during which the energy of electron transfer can be utilized (the B reduction or B oxidation stage), that has been carried on in the literature for many years, is rather meaningless. Energy utilization can only take place in the course of transformer conformational relaxation, which is not a redox process at all.

Figure 8.7 shows the energy changes of ABC system and those of individual carriers for each step of the process (8.39). Only relative energy changes being important in our discussion, the initial energy levels of all carriers, as well as those of the total system, are arbitrarily made to coincide.

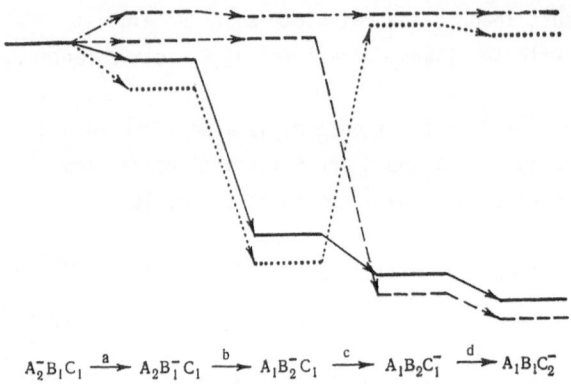

$$A_2^- B_1 C_1 \xrightarrow{a} A_2 B_1^- C_1 \xrightarrow{b} A_1 B_2^- C_1 \xrightarrow{c} A_1 B_2 C_1^- \xrightarrow{d} A_1 B_1 C_2^-$$

Fig.8.7. Scheme of changes in the total energy of ABC system (solid lines), of carriers A (dot-and-dash line), of carriers B (broken line), and of carriers C$^-$ (dotted line) during all stages of the process (8.39)

As long as the cooperativity of the membrane is not taken into account, we may neglect the intercarrier interactions. Therefore, the energy of the acceptor C does not change during stage "a". The energy of the donor A increases by the quantity practically equal to the ionization potential of A_2^- in the membrane minus the energy dissipating as a result of vibrational relaxation of A_2 active center and its adjacent environment (see Chaps.5 and 6). The energy of transformer B decreases by the value of B_1 electron affinity plus the corresponding dissipation energy. We can assume (though it is not obligatory) that this decrease is slightly greater than the increase of level A. Thus, the energy of the whole system decreases during stage "a". During stage "b" the level of C does not change, the level of A decreases slightly due to the conformational relaxation $A_2 \rightarrow A_1$, and the energy of transformer B undergoes a considerable decrease due to the slow conformational relaxation $B_1^- \rightarrow B_2^-$. The ability to undergo large energy changes in the course of conformational relaxation is a characteristic property of a transformer. Conformational relaxation results in the adjustment of the donor electron level of B_2^- active center[14] and the acceptor electron level of C_1 active center (with an accuracy of ~0.1 eV). It makes possible the electron tunnelling from B_2^- to C_1 with nonequilibrium states B_2 and C_1^- (stage "c") being formed. During this stage, the energy of A does not change, the energy of B increases to a value exceeding that of the initial equilibrium state B_1, and the energy of acceptor C undergoes a great decrease, mainly due to the relatively high electron affinity of C_1. It should be borne in mind that the difference between electron affinities of A_1 and C_1 (or between ionization potentials of A_2^- and C_2^-) is the main energy source of the net process.

---

[14]One must clearly distinguish between the electron levels and the levels of total carrier energy shown in Fig.8.7.

During the slow stage "d", the carriers B and C slowly relax to corresponding equilibrium conformations. Thus, as a result of the net process, the electron is transferred from A to C, these carriers now being in equilibrium conformations, and the transformer B is in the initial state after the cycle of redox, and conformational transformations has been completed.

According to this scheme, the "primary macroerg" is the transformer in state B_1^-, i.e., it has a reduced active center and is in the initial "oxidized", but now nonequilibrium, conformation. During the conformational relaxation of the transformer, the energy of this primary macroerg may be utilized for the synthesis of a chemical high-energy compound.

How can this be accomplished? One possible mechanisms has been suggested (but, surely, not proved) in [8.115]. There is a rather well-grounded opinion according to which in each coupling site there exists a labile complex between the transformer and the coupling enzyme active center, this complex containing adenine group of bound ADP. Conformational relaxation $B_1^- \rightarrow B_2^-$ affects not only the electronic level of B but also the acceptor level of adenine in such a way that during most of the time of relaxation these two levels will practically coincide (with the accuracy of ~0.1 eV). Under the conditions of the coinciding levels, the quantum-mechanical resonance ensures the dynamic delocalization of the electron [8.117]. Therefore, during slow adiabatic passage of electronic levels of the transformer and the adenine relative to each other, an excess of B^- electron density is extended to adenine.

The increase of electron density on the adenine ring (reduction of adenine) is accompanied by an increase of the basicity of its amino group, which can reach the basicity value characteristic of aliphatic amines [8.118].

Here, on the right, a reduced radical-anion form of adenine residue is shown. Its amino group has a heightened basicity and easily reacts with any electrophilic group. In [8.118] it has been assumed that the role of electrophilic group is played by the inorganic phosphate, and that covalent binding of phosphate to adenine amino group represents the first step of ATP formation. Schemes of this type agree, in principle, with the data obtained by KAYUSHIN et al. [8.119] on the appearance of free radicals during ATP enzymatic hydrolysis, and with the data obtained by MAKAROV et al. [8.120], according to which hydrogenated (reduced) ADP molecules can form a bond with phosphate.

It is, however, doubtful, that phosphate participates directly in the formation of the primary chemical high-energy compound [8.25]. The carboxyl scheme suggested in [8.121,122] can be considered as a working hypothesis. The increase of electron density on the adenine ring and the increase of amino group basicity led to the formation of an amide bond with one of the carboxyl groups of the coupling enzyme

For this state of adenine ring, the - N - C - bond is characterized by low free energy of hydrolysis, and, therefore, it is not a high-energy bond. However, the transformer relaxation proceeds further, and the electronic level of B becomes lower than that of the adenine. The electron density leaves adenine and returns to the B active center, the adenine amino group again becomes "aromatic", and the bond that has just been formed is now a high energy one

The discharge of the primary mechanical macroerg, B_1^-, and the formation of the primary chemical macroerg are, thus, realized during one elementary act of transformer conformational relaxation, and can, therefore, proceed without any considerable energy dissipation. It is of importance to understand that the decrease of electron energy level of the transformer is due not to electronic transition of any kind, but to the change in the shape of the potential well during relaxation $B_1^- \to B_2^-$. At any given moment of time the electron level remains the ground level. Excess energy of B_1^- cannot, therefore, be dissipated by means of intramolecular radiative or radiativeless electron or electron-vibrational transitions. The energy of B_1^- can be decreased only in the course of slow mechanical relaxation (see, however, later about the action of uncouplers).

So far, we have discussed the behavior of electron carriers without taking into account their membrane environment. The carriers and ATP synthesizing enzymes are, however, incorporated into the lipoprotein membrane, and the state of this membrane depends cooperatively on the states of its components. We have seen how the relaxation of the transformer strained conformation leads to the formation of the covalent high energy bond in ATP-synthetase. This local chemical change must, in turn, lead to the conformational relaxation of ATP-synthetase and its adjacent membrane environment. The new conformation of the local membrane sites is, thus, determined, in the end, by the carrier states and the high-energy bonds of primary chemical macroergs. We think that this new equilibrium conformation is equivalent to the so-called "energized state" of an energy transducing membrane [8.123], which is characterized by a higher solubility of paramagnetic hydrophobic probes [8.76], by a new level of fluorescent probe luminescence [8.124], etc. The membrane transition to the energized state is, therefore, a secondary process, and reflects the relaxation of the membrane to the new equilibrium conformation after the formation of the primary chemical macroerg. The characteristic times of membrane transitions are, accordingly, greater than those of the carriers [8.46].

It is important to understand that the energized state of a membrane is conformationally equilibrated, and, therefore, strictly speaking, is not energized (we may say that in this state the membrane is energized chemically, because it contains chemical macroergs). The energy stored in the membrane can be used in subsequent enzymatic acid-base reactions of ATP synthesis and in other endoergonic processes.

Figure 8.8 schematically shows the processes that take place at each coupling site of the membrane. Here, B_1SM represents the initial states of the transformer (B), ATP-synthetase (S) and the membrane (M). Due to electron addition (fast stage 1), the conformation of B does not change, but becomes strained and slowly relaxes to the equilibrium state B_2^- (stage 2). At the same time, the primary high-energy bond $(S \sim)$ is formed in the active center of S. Electron orbitals of B_2^- and of the next acceptor are now levelled, and the electron can be transferred along the chain (stage 3). The membrane, which has become strained due to the conformational change in S, relaxes to a new conformationally equilibrated state M^* (stage 4). In the state $B_2S \sim M^*$ (as well as in the state $B_2S \sim M$) transformer B cannot be reduced, because its unoccupied electron level lies much lower than that of the preceding donor (see Fig.8.7). In the absence of an energy acceptor (e.g., of ADP) the $B_2S \sim M^*$ state relaxes but slowly [8.125] to its initial equilibrium state. If ADP and P_i are present, the "chemical strain" will be released by ADP phosphorylation and $B_2S \sim M^*$ easily relaxes to the B_1SM state (stage 5) with reducible transformer B_1. Therefore, the rate of electron transfer through the coupling site is in this case determined by the rate of phosphorylation. Such a state of the membrane and the carriers corresponds to the so-called "respiratory control" state of mitochondria [8.25].

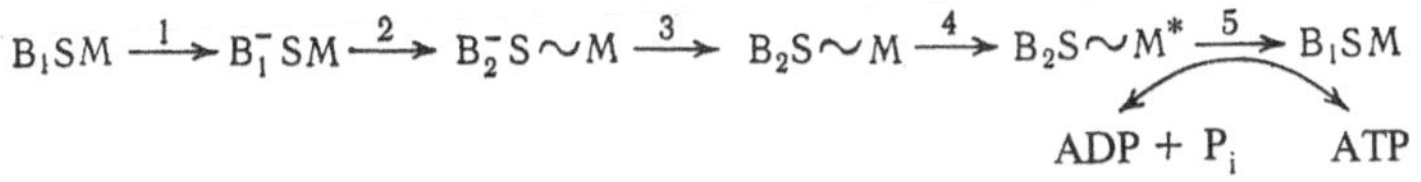

$$B_1SM \xrightarrow{1} B_1^-SM \xrightarrow{2} B_2^-S \sim M \xrightarrow{3} B_2S \sim M \xrightarrow{4} B_2S \sim M^* \xrightarrow{5} B_1SM$$

$$ADP + P_i \qquad ATP$$

Fig.8.8. Sequence of processes in a coupling site during membrane phosphorylation (a symplified scheme)

Due to the membrane cooperative properties the states of individual carriers and their relaxation times are interdependent. The initial state of the transformer before the working stroke (stage "b" in Fig.8.7) is essentially out of equilibrium, and the energy change during its relaxation is determined not only by the transformer characteristics but by the state of the fragments constituting the entire membrane, which, in turn, depends on the situation in other coupling sites in the same ETC, and, possibly, in other ETC's. The processes of this kind are, probably, responsible for the regulation of electron transport and energy transduction in membranes. A possibility is not excluded that this regulation automatically leads to such a state of every transformer at the moment of its reduction that the energy change during the relaxational transition is equal to or greater than that required for the synthesis of one ATP molecule. The process stoichiometry (passage of two electrons through a coupling site leads to the synthesis of one ATP molecule) is satisfied only on the average. The transfer of one electron through ETC from NADH to O_2 leads to the formation of one or two ATP molecules (on the average 1.5) depending on the momentary state of ETC. In a situation when the conformations of carriers and the positions of their electronic levels are essentially nonequilibrium, it is meaningless to use conventional thermodynamic functions, redox potentials included. The rates of certain stages of the cycle (see Fig.8.8) and, consequently, the ratio between the concentrations of the oxidized and the reduced forms of carriers may strongly depend on the membrane state. The paradoxical disagreement between the results of pulse and stationary measurements on mitochondrial ETC that has been mentioned above (see Chap.7) is, therefore, not surprising.

Relaxation to the equilibrium is possible not only as resulting from a macromolecule conformational change but also from the change of its surface charge. We can say that the conformation arising after the primary act (electron transfer in the case of a molecule, chemical macroerg formation in the case of a membrane) is out of equilibrium not only geometrically but also electrically. If a fast enough change of surface charge is possible, relaxation to the equilibrium state can then take place without any considerable geometrical changes. Perhaps, this is the way in which many uncouplers act. We have seen that, in the chemiosmotic concept, the uncoupling action of many compounds was ascribed to their ability to catalyze the transfer of hydrogen ions through the membrane with the consequent neutralization of its electric field. If uncouplers are indeed able to perform a fast electrical relaxation of individual carriers and of the whole membrane, this will result in the transformer electron levels changing so fast during this relaxation that the coupling system would not have enough time to synthesize a chemical macroerg, and the time of transfer through the coupling site would be diminished. Indeed, in the presence of uncouplers, the oxidation halftimes of corresponding electron carriers decrease by about an order of magnitude (see Sect.7.3).

The physical foundations of the membrane phosphorylation mechanism considered here are also in agreement with the work of KAHN [8.126], who has shown that there does not exist an intermediary chemical high-energy compound common for all the coupling sites.

The possible role of the membrane potential in the regulation of electron transport and membrane phosphorylation was discussed in [8.127].

In conclusion, it is necessary to emphasize again that the details of mechanisms considered here (e.g., the participation of adenine and ATP-synthetase carboxyl groups in the formation of primary chemical high-energy compounds) have been given only as an illustration of certain physical concepts.

It is quite obvious that effective energy transduction requires the excitation of such degrees of freedom which slowly exchange their energy with the thermal degrees of freedom. Mechanical degrees of freedom are, probably, the most appropriate for this purpose. We have seen that the ability to undergo directional conformational changes, i.e., to perform mechanical motion under the action of local perturbations is a distinctive feature of protein macromolecules. One can believe that this ability is used in all intracellular energy transformation processes, substrate and membrane phosphorylation included.[15]

It is possible that the required relaxational changes in ATP-synthetase can be induced without the participation of the transformer, but by means of other kinds of perturbations, e.g., by light (photoinduced transfer of an electron to the active center), or by sudden changes in pH, ionic composition, etc.

---

[15]Detailed discussions concerning the relaxational concept did appear recently [8.134,135].

9. Conclusion

In this book the author attempted to formulate certain general physical problems confronting a scientist who is trying to analyze biological phenomena.

This attempt was based on the notion of biological systems as a kind of machines, i.e., constructions possessing specific degrees of freedom and capable, on the one hand, of utilizing intrinsic energy to perform useful work, and, on the other hand, of creating a meaningful ordering, which is information. We have seen that these ideas are already applicable at the level of biopolymers.

Biological machines differ from those machines which can today be produced by engineers, first of all, by the fact that in biological machines the mechanical and statistical components are not separated, but coincide in space almost at all levels of organization.

Living systems, as all machines, are kinetically out-of-equilibrium. The pecularities of their functioning at various levels of organization are determined not so much by the fact that they are kept in an out-of-equilibrium state by energy and matter exchange with environment, as by the fact that transition to the equilibrium state is hindered kinetically.

The processes going on in living matter always obey all the laws of thermodynamics (either those of equilibrium or nonequilibrium, depending on the time intervals involved). However, due to the above-mentioned mechanical nature of living systems, the conventional thermodynamic approach is, in most cases, of small efficiency. The most interesting properties of living objects, even at the level of macromolecules and subcellular structures (such as enzymatic catalysis and energy transduction) are associated with the excitation of specific mechanical slowly relaxing degrees of freedom. I have tried (naturally, only phenomenologically) to outline and analyze the possible approaches to the solution of problems arising in this way.

Difficulties of making an adequate physical description of the basic characteristics of macromolecular and subcellular structures of living matter are connected not only with their kinetic out-of-equilibrium state, but also with the fact that this state changes during their functioning and due to rather subtle changes of such parameters as temperature or pH. This often makes it meaningless to utilize many ideas and equations, which have been successfully applied in the studies of simpler molecules. The development of quantitative and semi-empirical theories, capable of playing the role which Vant' Hoff's approach plays in thermodynamics and Arrhenius's approach in kinetics, represents one of the most important tasks of the science of biological physics.

For many biological phenomena it is as yet impossible to find not only the solution but even the formulation of the corresponding physical problems. A part of biology, which is so simple and so well explored that it is at least possible to formulate the physical questions related to it and to apply to it the principles of research developed in physics, can be called (rather unhappily) biological physics.

We can suppose that this fraction of biology will steadfastly grow approaching asymptotically the whole of biology (we are now only at the very beginning of this path). We cannot see any principal interdiction against it. One must not, however, forget that scientists today have no ideas about the possible physical approaches not only to such complicated biological phenomena as thinking, but to such a funda-mental and, consequently, simple process as morphogenesis. It was already said in Chap.1 that, to overcome these difficulties, it would scarcely be necessary to create a principally new, special "biological" physics. The only trouble is that our knowledge of biology is rather poor.

Life in the form in which it exists on our planet, and, in any case, the forms of life complex enough to possess the ability of self-knowledge, is, probably, a phenomenon not only extremely rare, but, I think, unique in the universe.

It would, therefore, be rather a pity not to avail ourselves of this opportunity and not to do our best to study living matter—the most interesting object of study for the living matter capable of studying.

References

An asterisk denotes a Russian publication, the title of which has been translated into English for the convenience of the reader.

Chapter 1

*1.1 L.A. Blumenfeld: In *On the Essence of Life* (Nauka, Moscow 1964) p.121
*1.2 M. Eigen: Naturwissenschaften *58*, 465 (1971)
1.3 N.V. Timofeev-Ressovskij, N.N. Voronzov, A.V. Jablokov: *The Brief Essay of Evolution Theory* (Nauka, Moscow 1969)

Chapter 2

2.1 E. Schrödinger: *What is life* (Cambridge University Press, Cambridge 1944)
2.2 C.E. Shannon, W. Weaver: *The Mathematical Theory of Communication* (University of Illionois Press, 1949)
2.3 J.C. Maxwell: *Theory of Heat* (London 1871)
2.4 L. Szillard: Z. Phys. *53*, 840 (1929)
2.5 P. Demers: Can. J. Res. A*22*, 27 (1944)
2.6 P. Demers: Can. J. Res. A*23*, 47 (1945)
2.7 J. Rothstein: Science *114*, 171 (1951)
2.8 J. Rothstein: Phys. Rev. *85*, 135 (1952)
2.9 L. Brillouin: *Science and Information Theory* (Academic, New York 1956)
2.10 G. Quastler: *The Rise of Biological Organization* (Yale University Press, New Haven 1964)

Chapter 3

*3.1 A. Chinchin: *The Mathematical Foundations of Statistical Mechanics* (Gostechisdat, Moscow 1943)
*3.2 A.M. Molchanov: In *Oscillatory Processes in Biological and Chemical Systems* (Nauka, Moscow 1967) p.292
3.3 L. Onsager: Phys. Rev. *37*, 405 (1931)
*3.4 L. Onsager: Phys. Rev. *38*, 2265 (1931)
*3.5 E.S. Bauer: *Theoretical Biology* (VIEM Press, Moscow 1935)
3.6 A.G. Goorwitch: *The Theory of Biological Field* (Nauka, Moscow 1944)
3.7 S.R. De Groot: *Thermodynamics of Irreversible Processes* (North-Holland, Amsterdam 1951)
3.8 S.R. De Groot, P. Mazur: *Nonequilibrium Thermodynamics* (North-Holland, Amsterdam 1962)
3.9 R. Haase: *Thermodynamik der irreversiblen Prozesse* (D. Steinkopf Verlag, Darmstadt 1963)
3.10 P.C. Glansdorff, J. Prigogine: Physica *30*, 351 (1964)
3.11 J. Prigogine: In *Theoretical Physics and Biology* (North-Holland, Amsterdam 1969) p.23
3.12 J. Keizer, R.F. Fox: Proc. Nat. Acad. Sci. USA *71*, 192 (1974)
3.13 A.J. Lotka: Z. Phys. Chem. *72*, 508 (1910)
3.14 A.J. Lotka: J. Am. Chem. Soc. *42*, 1595 (1920)
3.15 A.J. Lotka: *Elements of Physical Biology* (Dover, New York 1925)

3.16 V. Volterra: *Lecous sur la Théorie Mathematique de la Lutte pour la Vie* (Grauthiers-Villars, Paris 1931)

3.17 D.A. Frank-Kamenetzky: Dokl. Akad. Nauk SSSR *25*, 672 (1939)

*3.18 D.A. Frank-Kamenetzky: Uspechi Chimii *10*, 373 (1941)

3.19 B.P. Belousov: In *Synopses on Radiation Medicine for 1958* (Medgis, Moscow 1959) p.145

3.20 A.M. Zhabotinsky: Biofizika *9*, 306 (1964)

*3.21 A.M. Zhabotinsky: Dokl. Akad. Nauk SSSR *157*, 392 (1964)

3.22 A.M. Zhabotinsky: In *Oscillatory Processes in Biological and Chemical Systems* Nauka, Moscow 1967) p.149

*3.23 A.M. Zhabotinsky, A.N. Zaikin: In *Oscillatory Processes in Biological and Chemical Systems*, Vol.2 (Puschino-na-Oke, 1971) p.279

*3.24 A.M. Zhabotinsky, A.N. Zaikin: In *Oscillatory Processes in Biological and Chemical Systems*, Vol.2 (Pushino-na-Oke, 1971) p.288

*3.25 A.N. Zaikin, A.M. Zhabotinsky: Nature *225*, 535 (1970)

3.26 A.M. Zhabotinsky: *Auto-Oscillations of Concentrations* (Nauka, Moscow 1974)

3.27 *Biological Clocks*, Gold Spring Habor Symposia in quantitative biology, Vol.XXV, New York 1961)

*3.28 J.M. Romanovsky, H.V. Stepanova, D.S. Chernavsky: "What is mathematical bio-physics (*Kinetical Models in Biophysics*)" (Prosveschenije, Moscow 1971)

*3.29 E.E. Selkov: In *Oscillatory Processes in Biological and Chemical Systems* (Nauka, Moscow 1967) p.7

*3.30 E.E. Selkov: In *Oscillatory Processes in Biological and Chemical Systems*, Vol.2 (Puschino-na-Oke, 1971) p.5

3.31 N.M. Chernavskaja, D.S. Chernavsky: Biofizika *3*, 521 (1958)

*3.32 N.M. Chernavskaja, D.S. Chernavsky: UFN (Adv. in Phys.) *72*, 627 (1960)

3.33 D.S. Chernavsky, N.M. Chernavskaya: In *Oscillatory Processes in Biological and Chemical Systems* (Nauka, Moscow 1967) p.51

3.34 B. Hess, A. Boiteux: Ann. Rev. Biochem. *40*, 237 (1971)

3.35 L. Duysens, J. Amesz: Biochim. Biophys. Acta *24*, 19 (1957)

3.36 A.T. Wilson, M. Calvin: J. Am. Chem. Soc. *77*, 5948 (1955)

3.37 B. Chance, R.W. Estabrook, A. Ghosh: Proc. Nat. Acad. Sci. USA *51*, 1244 (1964)

3.38 B. Chance, B. Hess, A. Betz: Biochem. Biophys. Res. Commun. *16*, 182 (1964)

*3.39 B. Hess, B. Chance, A. Betz: Ber. Bunsenges. Phys. Chem. *68*, 718 (1964)

3.40 D.S. Chernavsky, J.N. Grigorov, M.S. Poljakova: In *Oscillatory Processes in Biological and Chemical Systems* (Nauka, Moscow 1967) p.138

3.41 E.E. Selkov: Eur. J. Biochem. , 79 (1968)

3.42 B.C. Goodwin: *Temporal Organization in Cells* (Academic Press, New York 1963)

Chapter 4

4.1 Yu.I. Churgin, D.S. Chernavsky, S.E. Schnoll: Mol. Biol. *1*, 419 (1967)

4.2 I.M. Lifshitz: J. Exp. Theor. Phys. *55*, 2408 (1968)

4.3 J.M. Lifshitz, A.Yu. Grosberg: J. Exp. Theor. Phys. *65*, 2399 (1973)

4.4 S.F. Edwards: Proc. Phys. Soc. *85*, 613 (1968)

*4.5 K.F. Freed: Adv. Chem. Phys. *22*, 1 (1972)

4.6 M.V. Volkenstein: "The Configurational Statistics of Polymer Chains", Publ. Ac. Sc. USSR (1959)

*4.7 J.M. Lifshitz, A.Yu. Grosberg: UFN (Adv. in Phys.) *113*, 331 (1974)

4.8 J.M. Lifshitz: Abstracts of papers, Symposia, IV Intern. Biophys. Congr., Moscow (1972) p.18

4.9 J.M. Lifshitz, A.Yu. Grosberg: Dokl. Akad. Nauk SSSR *220*, 468 (1975)

4.10 J.M. Lifshitz, A.Yu. Grosberg, A.R. Khokhlov: Rev. Mod. Phys. *50*, 683 (1978)

4.11 A.Yu. Grosberg: Biofizika *24*, 32 (1979)

Chapter 5

5.1 I.M. Lifshitz: J. Exp. Theor. Phys. *55*, 2408 (1968)
5.2 I.V. Berestetskaja, Yu.N. Kosaganov, Yu.S. Lazurkin, E.N. Trifonov, M.D. Frank-Kamenetskii: Mol. Biol. *4*, 137 (1970)
5.3 M.D. Frank-Kamenetskii: Dokl. Akad. Nauk SSSR *157*, 187 (1964)
5.4 M.D. Frank-Kamenetskii: Mol. Biol. *2*, 408 (1968)
5.5 D.M. Crothers, B.H. Zimm: J. Mol. Biol. *9*, 1 (1964)
5.6 Yu.S. Lazurkin, M.D. Frank-Kamenetskii, E.N. Trifonov: Biopolymers *9*, 1253 (1970)
5.7 B.H. Zimm: J. Chem. Phys. *33*, 1349 (1960)
5.8 M.D. Frank-Kamenetskii, Yu.S. Lazurkin: Ann. Rev. Bioph. Bioengin. *3*, 127 (1974)
5.9 M. Joly: *A Physico-Chemical Approach to the Denaturation of Proteins* (Academic, New York 1965)
5.10 W.W. Forest, J.M. Sturtevant: J. Am. Chem. Soc. *82*, 585 (1960)
5.11 P. Bro, J.M. Sturtevant: J. Am. Chem. Soc. *80*, 1789 (1958)
5.12 J. Steinhardt, E.M. Zaiser: J. Am. Chem. Soc. *75*, 1599 (1953)
5.13 E.M. Zaiser, J. Steinhardt: J. Am. Chem. Soc. *76*, 2866 (1954)
5.14 W.M. Jackson, J.F. Brandts: Biochemistry *9*, 2294 (1970)
5.15 P.L. Privalov, N.N. Khechilashvili, B.P. Atanasov: Biopolymers *10*, 1865 (1971)
5.16 P.L. Privalov: Abstracts of papers, Symposia, IV Intern. Biophysics Congress, Moscow (1972) p.16
5.17 P.L. Privalov: FEBS Lett. *40*, 140 (1974)
5.18 P.L. Privalov, N.N. Khechilashvili: J. Mol. Biol. *86*, 665 (1974)
*5.19 P.L. Privalov, N.N. Khechilashvili: Biofizika *19*, 14 (1974)
5.20 S.V. Konev, S.A. Aksentsev, E.A. Tshernitskii: *Cooperative Transitions of Proteins Within the Cell* (Nauka, Minsk 1970)
*5.21 E.A. Tshernitskii: *Luminescence and Structural Lability of Proteins in Solution and Within the Cell* (Nauka, Minsk 1972)
5.22 J.F. Brandts: J. Am. Chem. Soc. *86*, 4291 (1964)
5.23 J.F. Brandts: J. Am. Chem. Soc. *86*, 4302 (1964)
5.24 C. Tanford: Adv. Protein Chem. *23*, 121 (1968)
5.25 K. Okunuki: Adv. Enzymol. *23*, 29 (1961)
5.26 H.F. Epstein, A.N. Schlechter, R.F. Chen, C.B. Anfinsen: J. Mol. Biol. *60*, 499 (1971)
5.27 J.W. Teipel, D.E. Koshland, Jr.: Biochemistry *10*, 792 (1971)
5.28 J.W. Teipel, D.E. Koshland, Jr.: Biochemistry *10*, 798 (1971)
5.29 S. Glasstone, K.J. Laidler, H. Eyring: *Theory of Rate Processes: The Kinetics of Chemical Reactions, Viscosity, Diffusion, and Electrochemical Phenomena* (McGraw-Hill, New York 1941)
5.30 K.J. Laidler: *Chemical Kinetics*, 2nd ed. (McGraw-Hill, New York 1965)
5.31 E.C. Adams, M.R. Weiss: Biochem. J. *115*, 441 (1969)
5.32 F. Laviale, M. Rogart, A. Alfsen: Biochemistry *13*, 2231 (1974)
5.33 L.G. Clarke: Thesis, University Vermont (1966)
5.34 D.D.F. Shiao, J.M. Sturtevant: Biochemistry *8*, 4910 (1969)
5.35 C.H. Johnson, J.R. Knowles: Biochem. J. *101*, 56 (1966)
5.36 R.J. Foster, C. Niemann: J. Am. Chem. Soc. *77*, 3365 (1955)
5.37 A. Yapel: Thesis, University of Minnesota (1966)
5.38 D.D.F. Shiao: Biochemistry *9*, 1083 (1970)
5.39 H.J. Hinz, D.D.F. Shiao, J.M. Sturtevant: Biochemistry *10*, 1347 (1971)
5.40 S.F. Velick, J.P. Bagott, J.M. Sturtevant: Biochemistry *10*, 779 (1971)
5.41 M.W. Kirscher, H.K. Schachmann: Biochemistry *10*, 1900 (1971)
5.42 M.W. Kirscher, H.K. Schachmann: Biochemistry *10*, 1919 (1971)
5.43 G.I. Lichtenstein: Studia Biophysica *33*, 185 (1972)
*5.44 C.M. Dobson, R.J.P. Williams: FEBS Lett. *56*, 362 (1975)
5.45 L.A. Blumenfeld: *Hemoglobin and Reversible Attachment of Oxygen* (Sovjetskaja Nauka 1957)
5.46 M. Weissbluth: "The Physics of Hemoglobin", in *Structure and Bonding*, Vol.2 (Springer, Berlin, Heidelberg, New York 1967) p.1
5.47 M. Perutz: Nature *228*, 726 (1970)

5.48 M. Perutz: Nature *228*, 734 (1970)
5.49 L.A. Blumenfeld: Dokl. Akad. Nauk SSSR *85*, 1111 (1952)
5.50 J.L. Hoard: In *Hemes and Hemoproteins* (Academic, New York 1966) p.9
5.51 J.A. Hewitt, J.V. Kilmartin, L.F. Ten Eyck, M.F. Perutz: Proc. Nat. Acad. Sci. USA *69*, 203 (1972)
5.52 M. Perutz: Nature *237*, 495 (1972)
5.53 M. Perutz, L.F. Ten Eyck: Cold Spring Harbor Symp. Quant. Biol. *36*, 295 (1971)
5.54 A. Alpert, R. Banerjee, L. Lindgvist: Biochem. Biophys. Res. Commun. *46*, 913 (1972)
5.55 Q.H. Gibson: Biochem. J. *71*, 293 (1959)
5.56 A. Hvidt, S.O. Nielson: Adv. Protein Chem. *21*, 287 (1966)
5.57 S.W. Englander, N. Downer, H. Teitelbaum: Ann. Rev. Biochem. *41*, 903 (1972)
5.58 S.W. Englander, A. Rolfe: J. Biol. Chem. *248*, 4852 (1973)
5.59 L.A. Blumenfeld, R.A. Davydov, S.N. Magonov, R.O. Vilu: FEBS Lett. *49*, 246 (1974)
5.60 L.A. Blumenfeld, D.Sh. Burbajev, A.F. Vanin, R.O. Vilu, R.M. Davydov, S.N. Magonov: Zh. Strukt. Chim. *15*, 1030 (1974)
5.61 N.S. Fel', P.I. Dolin, R.M. Davydov, V.K. Vanag, S.P. Kuprin, N.M. Roldugina, L.A. Blumenfeld: Elektrokhimia *13*, 909 (1977)
5.62 G.S. Adair: J. Biol. Chem. *63*, 529 (1925)
5.63 R.T. Ogata, H.M. McConnell: Biochemistry *11*, 4792 (1972)
5.64 J. Monod, J. Wyman, J.P. Changeux: J. Mol. Biol. *12*, 88 (1965)
5.65 D.E. Koshland, Jr., G. Nemethy, D. Filmer: Biochemistry *5*, 365 (1966)
5.66 W.H. Huestis, M.A. Raftery: Biochemistry *14*, 1886 (1975)
5.67 Yu.A. Ermakov, V.I. Pasechnick, S.V. Tulskii: Biofizika (USSR) *20*, 591 (1975)
5.68 Yu.A. Ermakov, V.I. Pasechnick: Biofizika (USSR) *21*, 629 (1976)
5.69 Yu.A. Ermakov, S.P. Kuprin, V.I. Pasechnick: Biofizika (USSR) *21*, 788 (1976)
5.70 V.I. Pasechnick: Biofizika (USSR) *21*, 746 (1976)
5.71 L.A. Blumenfeld, Yu.A. Ermakov, V.I. Pasechnick: Biofizika (USSR) *22*, 8 (1977)
5.72 R. Dickerson, T. Takano, D. Eisenberg, O.B. Kallai, R. Swanson, A. Cooper, E. Margoliash: J. Biol. Chem. *246*, 1511 (1971)
5.73 T. Takano, R. Swanson, O.B. Kallai, R.E. Dickerson: Cold Spring Harbor Sympo. Quant. Biol. *36*, 397 (1971)
5.74 R. Swanson, B.L. Trus, N. Mandel, G. Mandel, O.B. Kallai, R.E. Dickerson: Preprint (1976)
5.75 T. Takano, B.L. Trus, N. Mandel, G. Mandel, O.B. Kallai, R. Swanson, R.E. Dickerson: Preprint (1976)
5.76 E. Margoliash, A. Schejter: Adv. Protein Chem. *21*, 113 (1966)
5.77 D.D. Ulmer, J.H.R. Käge: Biochemistry *7*, 2710 (1968)
5.78 L.A. Blumenfeld, R.M. Davydov, N.S. Fel, S.N. Magonov, R.O. Vilu: FEBS Lett. *45*, 256 (1974)
5.79 J. Pecht, M. Faraggi: Proc. Nat. Acad. Sci. USA *69*, 902 (1972)
5.80 I. Aviram, A. Schejter: Biopolymers *11*, 2141 (1972)
5.81 L.V. Belovolova, L.A. Blumenfeld, D.Sh. Burbajev, A.F. Vanin: Mol. Biol. *9*, 934 (1975)
5.82 S. Greschner, L.A. Blumenfeld, M. V. Genkin, R.M. Davydov, N.M. Roldugina: Stud. Biophys. *57*, 109 (1976)
5.83 L.A. Blumenfeld, S. Greschner, M.V. Genkin, R.M. Davydov, N.M. Roldugina: Stud. Biophys. *57*, 110 (1976)
5.84 G.I. Lichtenstein: Biofizika *11*, 24 (1966)
5.85 R. Lumry, Sh. Rajender: Biopolymers *9*, 1125 (1970)
5.86 G.I. Lichtenstein, B.I. Suchorukov: Zh. Fiz. Chim. *38*, 747 (1964)
5.87 J.G. Beetlestone, D.H. Irvine: J. Chem. Soc. A, 5090 (1964)
5.88 J.G. Beetlestone, D.H. Irvine: J. Chem. Soc. A, 3271 (1965)
5.89 A.C. Aunsiem, J.G. Beetlestone, D.H. Irvine: J. Chem. Soc. A, 357 (1966)
5.90 J.G. Beetlestone, D.H. Irvine: J. Chem. Soc. A, 951 (1968)
5.91 A.C. Aunsiem, J.G. Beetlestone, D.H. Irvine: J. Chem. Soc. A, 960 (1968)
5.92 P.L. Privalov: Thesis, Puschine-na-Oke (1970)
5.93 Ya.S. Lebedev, Yu.D. Tsvetkov, V.V. Vojevodskii: Kinet. Katal. *1*, 496 (1960)
*5.94 S.E. Shnoll: Vopr. Med. Chim. *4*, 443 (1958)
5.95 S.E. Shnoll: In *Molecular Biophysics* (Nauka, Moscow 1965) p.56

*5.96 S.E. Shnoll: In *Oscillatory Phenomena in Biological and Chemical Systems* (Nauka, Moscow 1967) p.22
*5.97 S.E. Shnoll: Thesis, Puschino-na-Oke (1970)
*5.98 S.E. Shnoll: In *Oscillatory Phenomena in Biological and Chemical Systems*, Vol.2 (Puschino-na-Oke, 1971) p.20
5.99 S.E. Shnoll, V.I. Grishina: Biofizika *9*, 376 (1964)
5.100 S.E. Shnoll, O.A. Rudneva, B.L. Nicolskaja, T.A. Revelskaja: Biofizika *6*, 166 (1961)
*5.101 E.P. Chetverikova: Biofizika *13*, 864 (1968)
*5.102 E.P. Chetverikova: In *Oscillatory Phenomena in Biological and Chemical Systems*, Vol.2 (Puschino-na-Oke, 1971) p.25
*5.103 V.V. Ribina, E.P. Chetverikova: In *Oscillatory Phenomena in Biological and Chemical Systems*, Vol.2 (Puschino-na-Oke, 1971) p.29
5.104 S.E. Shnoll, E.P. Chetverikova: Biochim. Biophys. Acta *403*, 89 (1975)
5.105 D. Meadows, O. Yardetzky: Proc. Nat. Acad. Sci. USA *61*, 406 (1968)
5.106 J.L. Markby, M.N. Williams, O. Yardetzky: Proc. Nat. Acad. Sci. USA *65*, 645 (1970)
5.107 H. Rüterjans, H. Witzel, O. Pongs: Biochem. Biophys. Res. Commun. *37*, 247 (1969)
5.108 K. Lindenström-Lang, J. Schellmann: *The Enzymes*, Vol.1 (Academic, New York 1959) p.443
5.109 L.A. Blumenfeld: In "Molecular Interactions and Activity in Proteins", Ciba Foundation Symp. 60 (New Series), Excerpta Med. London (1978) p.47
5.110 L.A. Blumenfeld, R.M. Davydov: Biochim. Biophys. Acta *549*, 255 (1979)
5.111 S.N. Magonov, L.A. Blumenfeld, V.K. Vanag, R.M. Davydov: Biofizika *23*, 414 (1978)
5.112 M.V. Genkin, R.M. Davydow, O.V. Krylov: Biofizika *23*, 29 (1978)
5.113 B.G. Westerman, R.M. Davydow, M.V. Genkin, L.A. Blumenfeld: Zh. Fiz. Khim. *53*, 1038 (1979)

Chapter 6

6.1 S.A. Berhard: *The Structure and Function of Enzymes* (Benjamin, New York 1968)
6.2 A.E. Braunstein: Z.V.Ch.O. im. Mendelejeva *8*, 81 (1963)
*6.3 M.V. Volkenstein: *The Physics of Enzymes* (Nauka, Moscow 1967)
6.4 M. Dixon, E.C. Webb: *Enzymes* (Academic, New York 1958)
6.5 M. Eigen, G.G. Hammes: Adv. Enzymol. *25*, 1 (1963)
6.6 S.E. Bresler: V Intern. Biochem. Congress, Symp. 1, Ed. Ac. Sc. USSR (1962) p.51
6.7 E.L. Smith, D.R. Cimmell, A. Light: V Intern. Biochem. Congress, Symp. 4, Ed. Ac. Sc. USSR (1962) p.148
6.8 J. Shevellier, J. Jacquot-Armand, J. Inn: Biochim. Biophys. Acta *92*, 521 (1964)
6.9 R. Frader, A. Light, E.L. Smith: J. Biol. Chem. *240*, 253 (1965)
6.10 E.M. Kosower: *Molecular Biochemistry* (McGraw-Hill, New York 1962)
6.11 D.E. Koshland, Jr., R.E. Neet: Ann. Rev. Biochem. *37*, 672 (1968)
6.12 D.E. Koshland, Jr.: J. Theor. Biol. *2*, 75 (1962)
6.13 S. Milstein, L.A. Cohen: Proc. Nat. Acad. Sci. USA *67*, 1143 (1970)
6.14 R. Lumry: *The Enzymes*, Vol.1 (Academic, New York 1959) p.157
6.15 H. Eyring, R. Lumry, J.D. Spikes: In *The Mechanism of Enzyme Action* (J. Hopkins University Press, Baltimore 1954) p.123
5.16 D.E. Koshland, Jr.: Proc. Nat. Acad. Sci. USA *44*, 98 (1958)
5.17 M.V. Volkenstein: Izv. Akad. Nauk SSSR, Biol. 805 (1971)
6.18 M.V. Volkenstein, R.R. Dogonadze, A.K. Madumarov, S.D. Urushadze, Yu.J. Charkaz: Mol. Biol. *6*, 431 (1972)
6.19 E.A. Moelwyn-Hughes: *The Enzymes*, Vol.1 (Academic, New York 1959) p.28
6.20 C.N. Hinshelwood: Proc. Roy. Soc. A*113*, 230 (1926)
6.21 L.S. Kassel: *The Kinetics of Homogeneous Gas Reactions* (The Chem. Catalog Comp., New York 1932)
6.22 N.B. Slater: Philos. Trans. Roy. Soc. A*246*, 57 (1953)

214

*6.23 E.S. Bauer: *Theoretical Biology* (VIEM Press, Moscow 1935)
6.24 N.I. Kobosev: Z. Fiz. Chim. *34*, 1443 (1960)
6.25 N.I. Kobosev: Z. Fiz. Chim. *21*, 1414 (1947)
*6.26 Yu.I. Churgin, D.S. Chernavsky, S.E. Shnoll: Mol. Biol. *1*, 419 (1967)
*6.27 S.E. Shnoll: In *Oscillatory Phenomena in Biological and Chemical Systems* (Nauka, Moscow 1967) p.22
6.28 V.I. Descherevsky, A.M. Zhabotinsky, E.E. Selkov, N.P. Sidorenko, S.E. Shnoll: Biofizika *15*, 225 (1970)
6.29 N.P. Sidorenko, V.I. Descherevsky: Biofizika *15*, 785 (1970)
6.30 F.J. Kajne, S.H. Suelter: J. Am. Chem. Soc. *87*, 897 (1965)
6.31 J. Reuben, F.J. Kajne: J. Biol. Chem. *246*, 6227 (1971)
6.32 G.G. Hammes, Cheng-Wen Wu: Science *172*, 1205 (1971)
6.33 W. Balthasar: Eur. J. Biochem. *22*, 158 (1971)
6.34 J. Schlessinger, A. Levitzki: J. Mol. Biol. *82*, 547 (1974)
6.35 K. Kirscher, E. Galleao, J. Schuster, D. Goodall: J. Mol. Biol. *58*, 29 (1971)
6.36 K. Kirscher, E. Galleao, J. Schuster, D. Goodall: J. Mol. Biol. *58*, 51 (1971)
6.37 G.I. Lichtenstein: Usp. Biol. Chim. *12*, 3 (1971)
6.38 G.I. Lichtenstein: Thesis, Moscow (1971)
6.39 G.I. Lichtenstein: Stud. Biophys. *33*, 185 (1972)
6.40 A.L. Buchachenko, A.M. Wasserman: Z. Strukt. Chim. *8*, 27 (1967)
6.41 A.N. Kusnetsov: Z. Strukt. Chim. *11*, 535 (1970)
6.42 V.P. Timofeev, O.L. Poljanovsky, M.V. Volkenstein, O.V. Preobrajenskaja, G.I. Lichtenstein, Yu. V. Kochanov: Mol. Biol. *6*, 377 (1972)
6.43 C.L. Hamilton, H.M. McConnell: In *Structural Chemistry and Molecular Biology* (W.H. Freeman, San Francisco 1968) p.115
6.44 H.M. McConnell, B.C. McFarland: Quart. Rev. Biophys. *3*, 91 (1970)
6.45 D.J. Stone, T. Buckman, P.L. Nordio, H.M. McConnell: Proc. Nat. Acad. Sci. USA *54*, 1010 (1965)
*6.46 A.N. Kusnetsov: Spin-Labels Method (Nauka, Moscow 1976)
6.47 G.I. Lichtenstein, Yu.D. Achmedov: Mol. Biol. *4*, 551 (1970)
6.48 G.I. Lichtenstein, V.P. Timofeev, M.V. Volkenstein, O.L. Poljanovsky, Yu.V. Kochanov: Mol. Biol. *5*, 1 (1971)
6.49 J.C. Seidel, J. Gergely, M. Chopek: Arch. Biochem. Biophys. *142*, 223 (1971)
6.50 L.G. Ignatjeva, T.M. Seregina, L.A. Blumenfeld, E.K. Rouge, R.I. Artjukh, G.P. Postnikova: Biofizika *17*, 533 (1972)
6.51 J.C. Seidel, J. Gergely: Arch. Biochem. Biophys. *158*, 853 (1973)
6.52 L.A. Blumenfeld, A.G. Ignatjeva: Eur. J. Biochem. *47*, 75 (1974)
6.53 N.V. Medvedeva, E.K. Runge, L.A. Blumenfeld: Biofizika *20*, 26 (1975)
6.54 E.K. Rounge, N.V. Medvedeva, M.N. Vilenkina, L.A. Blumenfeld: Biofizika *21*, 409 (1976)
6.55 L.A. Blumenfeld: Biofizika *16*, 724 (1971)
6.56 E. Goldberg, T. Wuntch: J. Exp. Zool. *165*, 101 (1967)
6.57 K.J. Laidler: *Chemical Kinetics of Enzyme Action* (Clarendon Press, Oxford 1958)
*6.58 S.V. Konev, S.A. Axentzev, E.A. Tshernitzky: *Cooperative Transitions of Proteins Within the Cell* (Nauka i technika, Minsk 1970)
6.59 R. Lumry, Sh. Rajender: Biopolymers *9*, 1125 (1970)
6.60 G. Talsky, W. Müller: H.S.Z. Physiol. Chem. *352*, 1681 (1971)
6.61 F.H. Johnson, H. Eyring, M.J. Polissar: *The Kinetic Basis of Molecular Biology* (Wiley, New York 1954)
6.62 J. Belehradek: *Temperature and Living Matter*, Protoplasm Monographien (Borntraeger-Verlag, Berlin 1935)
6.63 J. Belehradek: Ann. Rev. Physiol. *19*, 59 (1957)
6.64 A.W. Murray: Biochem. J. *103*, 271 (1967)
6.65 R.L. Miller, A.L. Bieber: Biochemistry *7*, 1420 (1968)
6.66 V. Massey, B. Curti, H. Ganther: J. Biol. Chem. *241*, 2347 (1966)
6.67 G.M. Lehrer, R. Barker: Biochemistry *9*, 1533 (1970)
6.68 M.R. Paule: Biochemistry *10*, 4509 (1971)
6.69 L.A. Blumenfeld: Biofizika *17*, 954 (1972)
6.70 A. Alpert, R. Bauerjee, L. Lindqvist: Biochem. Biophys. Res. Commun. *46*, 913 (1972)
6.71 V.N. Morozov: Biofizika *17*, 926 (1972)

6.72 L.A. Blumenfeld: J. Theor. Biol. *58*, 269 (1976)
6.73 N.V. Medvedeva, E.K. Rounge, L.A. Blumenfeld: Biochimija *39*, 999 (1974)
*6.74 D.S. Chernavskii: *Physical Models of Biological Catalysis* (Znanie, Moscow 1972)
6.75 B.I. Goldstein, M.A. Livshitz, M.V. Volkenstein: Mol. Biol. *8*, 784 (1974)
6.76 L.V. Belovolova, L.A. Blumenfeld, D.Sh. Burbajev, A.F. Vanin: Mol. Biol. *9*, 934 (1975)
6.77 A.E. Braunstein, M.G. Kritsman: Nature *140*, 503 (1937)
6.78 V.I. Ivanov, M.J. Karpeisky: Adv. Enzymol. *32*, 21 (1969)
6.79 Yu.A. Chismadzhev, V.F. Pastushenko, L.A. Blumenfeld: Biofizika *21*, 208 (1976)
6.80 V.M. Fain: J. Chem. Phys. *65*, 1854 (1976)
6.81 M.V. Volkenstein, I.B. Golovanov: Mol. Biol. *12*, 1377 (1978)
6.82 L.A. Blumenfeld, R.M. Davydov: Biochim. Biophys. Acta *549*, 255 (1979)
6.83 Yu.M. Romanovsky, N.K. Tikhomirova, Yu.I. Khurgin: Biofizika *24*, 442 (1979)

Chapter 7

7.1   L. Michaelis: J. Biol. Chem. *84*, 777 (1929)
7.2   L. Michaelis: J. Biol. Chem. *92*, 211 (1931)
7.3   L. Michaelis: J. Biol. Chem. *96*, 703 (1932)
7.4   E. Haas: Biochem. Z. *290*, 291 (1937)
7.5   B. Chance: In *Free Radical in Biological Systems* (Academic, New York 1961) p.1
7.6   B. Commoner, J. Townsend, G. Pake: Nature *174*, 689 (1954)
7.7   L.A. Blumenfeld: Izv. Akad. Nauk SSSR, Ser. Biol. 285 (1957)
7.8   L.A. Blumenfeld, A.E. Kalmanson: Dokl. Akad. Nauk SSSR *117*, 72 (1957)
7.9   L.P. Kajushin, I.K. Kolomijtseva, K.M. Lvov: Dokl. Akad. Nauk SSSR *134*, 1229 (1960)
7.10 M. Eigen, P. Matthies: Chem. Ber. *94*, 3309 (1961)
7.11 L.A. Blumenfeld, V.V. Vojevodsky, A.G. Semenov: *Electron Spin Resonance in Chemistry* (Wiley, New York 1973)
7.12 D.J.F. Ingram: *Free Radicals as Studied by Electron Spin Resonance* (Butterworth, London 1958)
7.13 *Free Radical in Biological Systems* (Academic, New York 1961)
7.14 *Magnetic Resonance in Biological Systems* (Pergamon Press, Oxford 1967)
7.15 B. Commoner, J.F. Ternberg: Proc. Nat. Acad. Sci. USA *47*, 1374 (1961)
7.16 J.R. Mallard, M. Kent: Nature *210*, 588 (1966)
7.17 O.N. Brzhevskaja, V.S. Marinov, O.S. Nedelina, E.M. Shekshejev: Biofizika *12*, 354 (1967)
7.18 G.A. Kernut, M.L. Edwards, K. Leech, K.A. Munday: Experientia *17*, 497 (1961)
7.19 A.V. Panemanglor: Thesis, Moscow (1975)
7.20 A.G. Chetverikov, L.A. Blumenfeld, G.V. Fomin: Biofizika *10*, 476 (1965)
7.21 L.A. Blumenfeld, A.G. Chetverikov, D.N. Kefalieva, A.F. Vanin: Stud. Biophys. *10*, 101 (1968)
7.22 A.G. Chetverikov: Biofizika *9*, 678 (1964)
7.23 A.G. Chetverikov, D.N. Kefalieva, A.F. Vanin: Biofizika *14*, 559 (1969)
7.24 O.N. Brzhevskaja, L.P. Kajushin, M.N. Kondrashova, O.S. Nedelina, E.M. Shekshejev: Biofizika *11*, 1076 (1966)
*7.25 E.K. Rounge, L.A. Blumenfeld: In *Bioenergetics and Biological Spectrophotometry* (Nauka, Moscow 1967) p.127
7.26 N.M. Emanuel, T.E. Lipatova: Dokl. Akad. Nauk SSSR *130*, 221 (1960)
7.27 A.E. Kalmanson, L.P. Lipchina, A.G. Chetverikov: Biofizika *6*, 410 (1961)
7.28 A.G. Chetverikov, A.E. Kalmanson, I.G. Charitonenkov, L.A. Blumenfeld: Biofizika *9*, 18 (1964)
7.29 I.G. Charitonenkov: Thesis, Moscow (1965)
7.30 A.E. Kalmanson, I.G. Charitonenkov, A.G. Chetverikov, L.A. Blumenfeld: Biofizika *8*, 722 (1963)
7.31 I.N. Naktinis, A.Ch. Chernjanskene: Biofizika *19*, 1039 (1974)
7.32 V.B. Golubev, M.N. Kuznetsova, V.B. Evdokimov: Zh. Fiz. Chim. *37*, 2795 (1963)
7.33 G.V. Fomin, L.A. Blumenfeld, B.I. Suchorukov: Dokl. Akad. Nauk SSSR *157*, 1199 (1964)

216

7.34   B.I. Liogonky, A.V. Ragimov, A.A. Berlin: Theor. Exp. Chim. *1*, 511 (1965)
7.35   B.I. Shapiro, V.M. Kazakova, Ya.K. Syrkin: Dokl. Akad. Nauk *165*, 619 (1965)
*7.36 G.V. Fomin, V.A. Rimskaja: In *Water State and Function in Biological Objects* (Nauka, Moscow 1967) p.131
7.37   S.D. Lazarov, A. Trifonov, T. Popov: Z. Phys. Chem. *238*, 145 (1968)
7.38   S.M. Shein, L.V. Bruchovetskaja, A.D. Chmelinskaja, V.F. Starichenko, T.M. Ivanova: Reakts. Sposobn. Org. Soedin. *6*, 1087 (1969)
7.39   S.M. Shein, L.V. Bruchovetskaja, F.V. Pishugin, V.F. Starichenko, V.N. Panfilov, V.V. Voejevodsky: Zh. Struct. Chim. *11*, 243 (1970)
*7.40 G.V. Fomin, L.A. Blumenfeld, R.M. Davidov, L.G. Ignatjeva: In *Water State and Function in Biological Objects* (Nauka, Moscow 1967) p.120
7.41   G.V. Fomin: Thesis, Moscow (1969)
7.42   V.B. Golubev, L.S. Yaguzhinsky, A.V. Volkov: Biofizika *11*, 572 (1966)
7.43   S.I. Sholina, G.V. Fomin, L.A. Blumenfeld: Zh. Fiz. Chim. *43*, 800 (1969)
7.44   L.A. Blumenfeld, L.V. Bruchovetskaja, G.V. Fomin, S.M. Shein: Zh. Fiz. Chim. *44*, 931 (1970)
7.45   G.V. Fomin, L.M. Gurgijan, L.A. Blumenfeld: Dokl. Akad. Nauk SSSR *191*, 151 (1970)
7.46   K.A. Bilevitch, O.Yu. Okhlobystin: Usp. Chim. *37*, 2162 (1968)
7.47   K.A. Bilevitch, N.N. Bubnov, O.Yu. Okhlobystin: Tetrahedron Lett. 3465 (1968)
*7.48 G.V. Fomin, L.A. Blumenfeld, L.M. Gurgijan: In "Macromolecular Compounds", ed. by Inst. of Chem. Phys. Ac. Sc. USSR (1970) p.284
7.49   G.V. Fomin, N.N. Katmazovsky, L.G. Ignatjeva, L.A. Blumenfeld: Biofizika *13*, 765 (1968)
7.50   L.M. Raichman, L.A. Blumenfeld: Biochimija *31*, 1127 (1966)
7.51   N.R. Boardman: Adv. Enzymol. *30*, 1 (1968)
7.52   G.M. Cheniae, J.F. Martin: Biochim. Biophys. Acta *197*, 219 (1970)
7.53   G. Hind, R.L. Heath, S. Izawa: In *Progress in Photosynthetic Resarch*, Vol.2 (Lichtenstein, 1969) p.1022
7.54   G. Vierke: Z. Naturforsch. *27b*, 172 (1972)
7.55   J.M. McCord, J. Fridovich: J. Biol. Chem. *243*, 5753 (1968)
7.56   J.M. McCord, J. Fridovich: J. Biol. Chem. *244*, 6049 (1969)
7.57   B.B. Keele, J.M. McCord, J. Fridovich: J. Biol. Chem. *245*, 6176 (1970)
7.58   G. Rotilio, L. Calabrese: Biochemistry *11*, 2182 (1972)
7.59   H. Metsner: H.S.Z. Physiol. Chem. *349*, 1586 (1968)
7.60   G.G. Komissarov, Yu.S. Shumov: Dokl. Akad. Nauk SSSR *182*, 1226 (1968)
7.61   G.G. Komissarov, A.N. Asanova, Yu.S. Shumov: Dokl. Akad. Nauk SSSR *206*, 1468 (1972)
7.62   G.G. Komissarov, Yu.S. Shumov, Yu.E. Borisevich: Dokl. Akad. Nauk SSSR *187*, 670 (1969)
7.63   A.A. Krasnovsky, G.P. Brin: Dokl. Akad. Nauk SSSR *139*, 142 (1961)
7.64   G.V. Fomin, G.P. Brin, M.V. Genkin, A.K. Lubimova, L.A. Blumenfeld, A.A. Krasnovsky: Dokl. Akad. Nauk SSSR *212*, 424 (1973)
7.65   L.A. Blumenfeld, M.I. Temkin: Biofizika *7*, 731 (1962)
7.66   O.N. Brzhevskaja, K.M. Lvov, O.S. Nedelina: Biofizika *9*, 500 (1964)
7.67   O.N. Brzhevskaja, L.P. Kajushin, O.S. Nedelina: Biofizika *11*, 213 (1966)
7.68   O.N. Brzhevskaja, O.S. Nedelina: Biofizika *13*, 141 (1968)
7.69   B. Chance: FEBS Lett. *23*, 3 (1972)
7.70   B.S.S. Masters, N.H. Bilimoria, H. Kamin, Q.H. Gibson: J. Biol. Chem. *240*, 4081 (1965)
7.71   R.C. Bray, P.F. Knowles, L.S. Meriwetter: In *Magnetic Resonance in Biological Systems* (Pergamon Press, Oxford 1967) p.249
7.72   P.T. Knowles: In *Magnetic Resonance in Biological Systems* (Pergamon Press, Oxford 1967) p.265
7.73   J.T. Spence, M. Heydanek, P. Hemmerich: In *Magnetic Resonance in Biological Systems* (Pergamon Press, Oxford 1967) p.269
7.74   E. Racker: *Mechanisms in Bioenergetics* (Academic, New York 1965)
7.75   S.P.J. Albrecht, E.C. Slater: Biochim. Biophys. Acta *245*, 503 (1971)
7.76   S.P.J. Albrecht, E.C. Slater: Biochim. Biophys. Acta *245*, 508 (1971)
7.77   R.A. Glegg, P.B. Garland: Biochem. J. *124*, 135 (1971)
7.78   M. Gutman, T.P. Singer, H. Beinert: Biochem. Biophys. Res. Commun. *44*, 1572 (1971)

7.79 N.R. Orme-Johnson, W.H. Orme-Johnson, R.E. Hansen, H. Beinert, J. Hatefi: Biochem. Biophys. Res. Commun. *44*, 446 (1971)
7.80 E.C. Slater: Quart. Rev. Biophys. *4*, 35 (1971)
7.81 P.B. Garland: Biochem. J. *118*, 329 (1970)
7.82 D.O. Hall, R. Cammack, K.K. Rao: In *Iron in Biochemistry and Medicine*, ed. by A. Jacob, M. Worwood (Academic, London 1974) p.279
7.83 D.Sh. Burbajev, A.V. Lebanidze: Biofizika *21*, 942 (1976)
7.84 D.E. Green, S. Fleischer: In *Horizons in Biochemistry* (Academic, New York 1962) p.381
7.85 M. Klingenberg, A. Kröger: In *Biochemistry of Mitochondria* (Academic, New York 1967) p.11
7.86 B. Chance, E.R. Redfearn: Biochem. J. *80*, 632 (1961)
7.87 B. Chance, D.F. Wilson, P.L. Dutton, M. Erecinska: Proc. Nat. Acad. Sci. USA *66*, 1175 (1970)
7.88 H. Theorell, A. Akesson: Science *90*, 67 (1939)
7.89 A. Schejter, P. George: Biochemistry *3*, 1045 (1964)
7.90 G. Greenwood, G. Palmer: J. Biol. Chem. *240*, 3660 (1965)
7.91 G. Brandt, P.C. Parks, G.H. Czerlinski: J. Biol. Chem. *241*, 4180 (1966)
7.92 G. Czerlinski, V. Bracokova: Arch. Biochem. Biophys. *147*, 707 (1971)
7.93 M.F. Wilson, C. Greenwood: Eur. J. Biochem. *22*, 11 (1971)
7.94 B. Chance, C. Lee, L. Mela: In Abstr. Symp. on Cytochromes, Osaka (1967) p.181
7.95 K.S. Amble, A. Venkataraman: Biochem. Biophys. Res. Commun. *1*, 133 (1959)
7.96 R.C. Criddle, R.M. Bock: Biochem. Biophys. Res. Commun. *1*, 138 (1959)
7.97 B. Chance, M. Erecinska: Arch. Biochem. Biophys. *143*, 675 (1971)
7.98 H. Beinert, D.E. Griffiths, D.C. Wharton, R.H. Sands: J. Biol. Chem. *237*, 2337 (1962)
7.99 M. Morrison, S. Horie, H.S. Mason: J. Biol. Chem. *238*, 2220 (1963)
7.100 *Oxidases and Related Redox Systems* (Wiley, New York 1965) p.929
7.101 R.C. Criddle, R.M. Bock, D.E. Green, H. Tisdale: Biochemistry *1*, 827 (1962)
7.102 B. Chance, W.D. Bonner, Jr., B.T. Storey: Ann. Rev. Plant Physiol. *19*, 295 (1968)
7.103 B. Chance: Nature *169*, 215 (1952)
7.104 B. Chance, R. Ernster, P.B. Garland, C.P. Lee, P.A. Light, T. Ohnishi, C.J. Ragan, D. Wang: Proc. Nat. Acad. Sci. USA *57*, 1498 (1967)
7.105 D.E. Fleischmann, G. Tollin: Biochim. Biophys. Acta *94*, 248 (1965)
*7.106 V.P. Skulachev: *Energy Accumulation Within the Cell* (Nauka, Moscow 1969)
7.107 D.F. Wilson, P.L. Dutton: Arch. Biochem. Biophys. *136*, 583 (1970)
7.108 J.S. Rieske: Arch. Biochem. Biophys. *145*, 179 (1971)
7.109 B. Chance, D. De Vault, V. Legalais, L. Mela, T. Jonetani: In *Fast Reactions and Primary Processes in Chemical Kinetics*, 5th Nobel Symp. (Almqvist and Wiksell, Stockholm 1967) p.437
7.110 B. Chance, C.P. Lee, T. Ohnishi, J. Higgins: In *Electron Transport and Energy Conservation* (Adriatica Editrice, Bari 1970) p.29
7.111 B. Chance, B. Shoener, D. De Vault: In *Oxidases and Related Redox Systems*, ed. by M.R. Ring, H.S. Mason, M. Moorison (Wiley, New York 1965), p.907
7.112 J. Pecht, M. Faraggi: Proc. Nat. Acad. Sci. USA *69*, 902 (1972)
7.113 M.V. Genkin, R.M. Davidov, O.V. Krylov, L.A. Blumenfeld: Dokl. Akad. Nauk SSSR *232*, 367 (1977)
7.114 L.P. Vernon: Bacteriol. Rev. *32*, 243 (1968)
7.115 J.S. Leigh, Jr., P.L. Dutton: Biochem. Biophys. Res. Commun. *46*, 414 (1972)
7.116 D.L. Keistler: J. Biol. Chem. *240*, 2673 (1965)
7.117 D.B. Knaff, D.I. Arnon: Proc. Nat. Acad. Sci. USA *64*, 715 (1969)
7.118 U. Siggel, G. Renger, H.H. Stiel, B. Rumberg: Biochim. Biophys. Acta *256*, 328 (1972)
7.119 S.V. Hangulov, M.G. Goldfeld, L.A. Blumenfeld: Dokl. Akad. Nauk SSSR *218*, 726 (1974)
7.120 S.V. Hangulov, M.G. Goldfeld, L.A. Blumenfeld: Stud. Biophys. *48*, 85 (1975)
7.121 M.G. Goldfeld, S.V. Hangulov, L.A. Blumenfeld: Photosynthetica *12*, 21 (1978)
7.122 B. Chance, M. Nishimura: Proc. Nat. Acad. Sci. USA *46*, 19 (1960)
7.123 H.T. Witt, A. Müller, B. Rumberg: Nature *192*, 967 (1961)
7.124 B. Chance, W.D. Bonner, Jr.: In "Photosynthetic Mechanisms of Green Plants", Publication No. 1145, Nat. Acad. Sci. USA, Washington (1963) p.66

7.125 W.J. Wredenberg, L.N.M. Duysens: Biochim. Biophys. Acta *79*, 456 (1964)
7.126 D. De Vault, B. Chance: Biophys. J. *6*, 825 (1966)
7.127 D. De Vault, J.H. Parks, B. Chance: Nature *215*, 642 (1967)
7.128 P.L. Dutton, T. Kihara, B. Chance: Arch. Biochem. Biophys. *139*, 236 (1970)
7.129 L.N. Grigorov: Thesis, Moscow University (1970)
7.130 L.N. Grigorov, A.A. Kononeko, A.B. Rubin: Mol. Biol. *4*, 483 (1970)
7.131 L.N. Grigorov, D.S. Chernavsky: Biofizika *17*, 195 (1972)
7.132 D.S. Bendall, D. Sofroya: Biochim. Biophys. Acta *234*, 371 (1971)
7.133 C.S. Jang, W.E. Blumberg: Biochem. Biophys. Res. Commun. *46*, 422 (1972)
7.134 P. Jordan: Naturwissenschaften *26*, 693 (1938)
7.135 A. Szent-Gyorgyi: Science *93*, 609 (1941)
7.136 A. Szent-Gyorgyi: Nature *148*, 157 (1941)
7.137 Yu.A. Vladimirov, S.V. Konev: Biofizika *2*, 3 (1957)
7.138 Yu.A. Vladimirov, S.V. Konev: Biofizika *4*, 533 (1959)
7.139 G. Weber: Biochem J. *75*, 345 (1960)
*7.140 S.V. Konev: *Electronic Excitation States of Biopolymer* (Ed. Akad. Nauk BSSR,
* Minsk 1965)
7.141 Yu.A. Vladimirov: *Photochemistry and Luminescence of Proteins* (Nauka,
Moscow 1965)
*7.142 S.I. Vavilov: *Microstructure of Light* (Ed. Akad. Nauk SSSR, Minsk 1950)
7.143 T. Förster: Ann. Phys. *2*, 55 (1948)
7.144 A.N. Terenin: Izv. Akad. Nauk SSSR, Ser. Biol. 369 (1947)
7.145 A.N. Terenin, A.A. Krasnovsky: Usp. Fiz. Nauk *37*, 65 (1949)
7.146 M.G. Evans, J. Gergely: Biochim. Biophys. Acta *3*, 188 (1949)
7.147 L.A. Blumenfeld: J. Exp. Theor. Phys. *18*, 670 (1948)
7.148 L.E. Lyons: J. Chem. Soc. 5001 (1957)
7.149 V.A. Bendersky, L.A. Blumenfeld: Dokl. Akad. Nauk SSSR *144*, 813 (1962)
7.150 L.A. Blumenfeld, V.A. Bendersky: Zh. Struct. Chim. *4*, 405 (1963)
7.151 V.A. Bendersky, L.A. Blumenfeld, D.A. Popov: Zh. Struct. Chim. *7*, 370 (1966)
7.152 L.A. Blumenfeld, V.A. Bendersky, P.A. Stunzhas: Zh. Struct. Chim. *7*, 686 (1966)
7.153 F. Gutman, L.E. Lyons: *Organic Semiconductors* (Wiley, New York 1967)
7.154 E.J. Murphy, A.C. Walner: J. Phys. Chem. *32*, 1761 (1928)
7.155 G. King, J.A. Medley: J. Colloid Sci. *4*, 1 (1949)
7.156 M.H. Cardew, D.D. Eley: Disc. Faraday Soc. *27*, 115 (1959)
7.157 D.D. Eley: In *Horizons in Biochemistry* (Academic, New York 1962) p.341
7.158 C.T.O'Konski, P. Moser, M. Shirai: In *Quantum Aspects of Polypetides and
Polynucleotides* (Interscience, New York 1964) p.479
*7.159 L.I. Boguslavsky, A.V. Vannikov: *Organic Semiconductors* (Nauka, Moscow 1968)
*7.160 *Organic Semiconductors*, ed. by V.A. Kargin (Nauka, Moscow 1968)
*7.161 A.A. Dulov, A.A. Slinkin: *Organic Semiconductors* (Nauka, Moscow 1970)
7.162 E.M. Trukhan: Fiz. Tverd. Tela *4*, 3496 (1962)
7.163 E.M. Trukhan: Biofizika *11*, 412 (1966)
7.164 L.A. Blumenfeld, D.N. Kafalijeva, V.A. Livshitz, I.S. Solovjev, A.G.
Chetverikov: Dokl. Akad. Nauk SSSR *193*, 700 (1970)
7.165 D.D. Eley, R. Pethig: J. Bioenerg. *2*, 39 (1971)
7.166 D.D. Eley, R. Pethig: Disc. Faraday Soc. *51*, 164 (1971)
7.167 L.A. Blumenfeld, O.P. Samoilova, I.S. Solov'yev: Biofizika *19*, 87 (1974)
7.168 O.P. Samoylova, I.S. Solovyev, L.A. Blumenfeld: Biofizika *21*, 1031 (1976)
7.169 I.S. Solovyev, O.P. Samoylova, L.A. Blumenfeld: Biofizika *21*, 1035 (1976)
7.170 V.A. Livshitz: Thesis, Moscow (1969)
7.171 L.A. Blumenfeld: IV Intern. Biophys. Congr., Reports, Symposiums, Moscow
(1972) p.56
7.172 M.K. Pulatova, V.N. Rogumenkova, L.P. Kayushin: Biofizika *6*, 548 (1961)
7.173 M.K. Pulatova, O.A. Asisova: Biofizika *9*, 33 (1964)
*7.174 L.P. Kayushin, K.M. Lvov, M.K. Pulatova: *Investigation of Paramagnetic
Centers of Irradiated Proteins* (Nauka, Moscow 1970)
7.175 E.N. Sudbina, M.K. Pulatova, L.P. Kayushin: Biofizika *16*, 596 (1971)
7.176 N.E. Geacintov, F. Van Nostrand, J.F. Becker, J.B. Tinnel: Biochim. Biophys.
Acta *267*, 65 (1972)
7.177 F.V. Nostrand, N.E. Geacintov, J.F. Becker: In Book of Abstracts, VI Intern.
Congr. on Photobiology, Bochum (1972) No.230
7.178 A.Yu. Borisov, M.D. Iljina: Biochimia *36*, 822 (1971)

7.179 A.N. Terenin, E.K. Putseiko: V Intern. Bioch. Congr. Symp. 6, (Ed. Akad. Nauk SSSR, Minsk 1962) p.52
7.180 R.J. Cherry: Quart. Rev. 22, 160 (1968)
7.181 I.S. Meylanov, V.A. Bendersky, L.A. Blumenfeld: Biofizika 15, 822 (1970)
7.182 I.S. Meglanov, V.A. Bendersky, L.A. Blumenfeld: Biofizika 15, 959 (1970)
7.183 E.C. Weaver: Ann. Rev. Plant Physiol. 19, 283 (1968)
7.184 E.C. Weaver: Nature 226, 183 (1970)
*7.185 V.A. Bendersky, L.A. Blumenfeld, A.I. Pristupa: Vysokomol. Soedin. (Highmol. comp.) A9, 171 (1967)
7.186 E.S. Cherepanova: Thesis, Kazan University (1973)
7.187 H. Beinert, B. Kok: In "Photosynthetic Mechanisms of Green Plants", Publ. 1145, Nat. Acad. Sci. USA, Washington (1963) p.131
7.188 T. Hiyama, B. Ke: Biochim. Biophys. Acta 267, 160 (1972)
7.189 D.N. Kafalieva, L.A. Blumenfeld, I.S. Soloviev, V.A. Livshitz, A.P. Darmanjan: Biofizika 14, 1117 (1969)
7.190 A. Kerkeni, A.T. Chernokolev, A.K. Kukushkin, M.K. Solntzev: In "Theor. Exp. Biofizika", ed. by Kaliningr. University (1973) No.4, p.91
7.191 A.K. Kukushkin: Biofizika 13, 1124 (1968)
7.192 F. Gutmann: Nature 219, 1359 (1968)
7.193 L.A. Blumenfeld, D.S. Chernavskii: J. Theor. Biol. 39, 3 (1973)
7.194 L.J. Osterhoff, A. Kuppermann: In "Fast Reactions and Primary Processes in Chemical Kinetics" (5th Nobel Symp.) Stockholm (1967) p.466
7.195 L.A. Blumenfeld, V.I. Goldansky, M.I. Podgoretzky, D.S. Chernavskii: Zh. Strukt. Chimii 8, 854 (1967)
7.196 J.H. Callomon, T.M. Dunn, J.M. Mills: Phil. Trans. Roy. Soc. A259, 499 (1966)
7.197 V.G. Levich, A.M. Kuznetzov: Theor. Exp. Chimija 6, 291 (1970)
7.198 R.R. Dogonadze, A.M. Kuznetzov: Theor. Exp. Chimija 6, 298 (1970)
7.199 V.G. Levich, R.R. Dogonadze, M.A. Vorotyntzev, E.D. German, A.M. Kuznetzov, Yu.I. Harkatz: Electrochimiya 6, 562 (1970)
7.200 D. DeVault: J. Theor. Biol. 62, 115 (1976)
7.201 B.G. Westerman, R.M. Davydov, M.V. Genkin, L.A. Blumenfeld: Zh. Fiz. Khim. 53, 1038 (1979)
7.202 M.G. Goldfeld, L.A. Blumenfeld: Bull. Magnetic Resonance 1, 66 (1979)

Chapter 8

8.1 F. Lipmann: Adv. Enzymol. 1, 99 (1941)
8.2 E. Racker: Mechanisms in Bioenergetics (Academic, New York 1965)
*8.3 V.P. Sculachev: Energy Accumulation Within the Cell (Nauka, Moscow 1969)
8.4 H. Kalckar: Biological Phosphorylation. Developments of Concepts (Prentice-Hall, Englewood Cliffs, New York 1969)
8.5 D. Stellen: Amer. J. Med. 28, 867 (1960)
8.6 K. Shikama: Arch. Biochem. Biophys. 147, 311 (1971)
8.7 J. Rosing, E.C. Slater: Biochim. Biophys. Acta 267, 275 (1972)
8.8 R.A. Alberty: J. Biol. Chem. 243, 1337 (1968)
8.9 R.A. Alberty: J. Biol. Chem. 244, 3290 (1969)
8.10 T. Benzinger, C. Kitzinger, R. Hems, K. Burton: Biochem. J. 71, 400 (1959)
8.11 B.E.C. Banks, C.A. Vernon: J. Theor. Biol. 29, 301 (1970)
8.12 R.J. Gillespie, G.A. Maw, C.A. Vernon: Nature 171, 1147 (1953)
8.13 B.E. Banks: Chem. Br. 5, 514 (1969)
8.14 R.A. Ross, C.A. Vernon: Chem. Br. 6, 539 (1970)
8.15 B.E.C. Banks, C.A. Vernon: Chem. Br. 6, 541 (1970)
8.16 L. Pauling: Chem. Br. 6, 468 (1970)
8.17 D. Wilkie: Chem. Br. 6, 472 (1970)
8.18 A.F. Huxley: Chem. Br. 6, 477 (1970)
8.19 C.W.F. McClare: J. Theor. Biol. 35, 233 (1972)
8.20 T.L. Hill, M.T. Morales: J. Am. Chem. Soc. 73, 1656 (1951)
8.21 A. Pullman, B. Pullman: Quantum Biochemistry (Wiley Interscience, New York 1963)
8.22 H. Kalckar: Biol. Rev. Cambr. Phil. Soc. 17, 28 (1942)

8.23 R.E. Alving, K. Laki: J. Theor. Biol. *34*, 199 (1972)
8.24 C.W.F. McClare: J. Theor. Biol. *30*, 1 (1971)
8.25 A.L. Lehminger: *The Mitochondria* (Benjamin, New York 1964)
8.26 W.A. Engelhardt: Biochem. Z. *227*, 16 (1930)
8.27 W.A. Engelhardt: Biochem. Z. *251*, 343 (1932)
8.28 H. Kalckar: Enzymologia *2*, 47 (1937)
8.29 H. Kalckar: Biochem. J. *33*, 631 (1939)
8.30 V.A. Belitzer, E.T. Cybakova: Biochimia *4*, 516 (1939)
8.31 S. Ochoa: J. Biol. Chem. *138*, 751 (1941)
8.32 S. Ochoa: J. Biol. Chem. *151*, 493 (1943)
8.33 B. Chance, G.R. Williams: Nature *176*, 250 (1955)
8.34 B. Chance, G.R. Williams: J. Biol. Chem. *217*, 429 (1955)
8.35 B. Chance, G. Hollinger: J. Am. Chem. Soc. *79*, 2970 (1957)
8.36 B. Chance, G.R. Williams: Adv. Enzymol. *17*, 65 (1965)
8.37 E.C. Slater: Quart. Rev. Biophys. *4*, 35 (1971)
8.38 S. Muraoka, E.C. Slater: Biochim. Biophys. Acta *180*, 221 (1969)
8.39 R.S. Cockrell, E.J. Harris, B.C. Pressman: Biochemistry *5*, 2326 (1966)
8.40 E.C. Slater: In *Electron Transport and Energy Conservation* (Adriatica
      Editrice, Bari 1970) p.363
8.41 B. Chance: FEBS Lett. *23*, 3 (1972)
8.42 D.F. Wilson, M. Erecinska, T. Ohnishi, P.L. Dutton: Bioelectrochem. Bioener-
      getics *1*, 3 (1974)
8.43 M. Klingenberg: In *Biological Oxidations* (Interscience, New York 1968) p.3
8.44 D.F. Wilson, P.L. Dutton: Arch. Biochem. Biophys. *136*, 583 (1970)
8.45 D.F. Wilson, P.L. Dutton: Biochem. Biophys. Res. Commun. *39*, 59 (1970)
8.46 B. Chance, D.F. Wilson, P.L. Dutton, M. Erecinska: Proc. Nat. Acad. Sci.
      USA *66*, 1175 (1970)
8.47 J.S. Rieske: Arch. Biochem. Biophys. *145*, 179 (1971)
8.48 J.A. Berden, F.P. Opperdoes: Biochim. Biophys. Acta *267*, 17 (1972)
8.49 P.L. Dutton, M. Erecinska, N. Sato, J. Mukai, D.T. Wilson: Biochim. Biophys.
      Acta *267*, 15 (1972)
8.50 B. Chance, M. Erecinska: Arch. Biochem. Biophys. *143*, 675 (1971)
8.51 A.H. Caswell: Arch. Biochem. Biophys. *144*, 445 (1971)
8.52 D.D. Tyler, R.W. Eastabrook, D.R. Sanadi: Arch. Biochem. Biophys. *114*, 239
      (1966)
8.53 P. Joliot, A. Joliot, B. Kok: Biochim. Biophys. Acta *153*, 635 (1968)
8.54 A.B. Rubin, T.E. Krendeleva: Usp. Sovr. Biol. *73*, 364 (1972)
8.55 M. Avron, B. Chance: In *Currents in Photosynthesis* (Denker, Rotterdam 1966)
      p.455
8.56 G.A. Hauska, R.E. McCarty, E. Racker: Biochim. Biophys. Acta *197*, 206 (1970)
8.57 A.T. Jagendarf, M. Marulies: Arch. Biochem. Biophys. *90*, 184 (1960)
8.58 H. Bohme, W.A. Cramer: Biochemistry *11*, 1155 (1972)
8.59 S. Izawa, N.E. Good: Biochim. Biophys. Acta *162*, 380 (1968)
8.60 S. Saha, R. Quitrakul, S. Izawa, N.E. Good: J. Biol. Chem. *246*, 3201 (1971)
8.61 S.C. Reeves, P. Heathcole, D.O. Hall: In "Book of Abstracts, VI Intern.
      Congr. on Photobiology", Bochum (1972), No.284
8.62 H. Böhme, A. Trebst: Biochim. Biophys. Acta *180*, 137 (1969)
8.63 E. Racker, N. Nelson, D. Deters: In "Book of Abstracts, VI Intern. Congr.
      on Photobiology", Bochum (1972), No.003
8.64 E.C. Slater: Nature *172*, 975 (1953)
8.65 T. King, M. Kubojama, S. Takemori: In *Oxidases and Related Redox Systems*,
      Vol.2 (Wiley Interscience, New York 1965) p.707
8.66 J.T. Penniston, R.A. Harris, J. Asai, D.E. Green: Proc. Nat. Acad. Sci. USA
      59, 624 (1968)
8.67 R.A. Harris, J.T. Penniston, J. Asai, D.E. Green: Proc. Nat. Acad. Sci. USA
      59, 830 (1968)
8.68 P. Mitchell: Nature *191*, 144 (1961)
8.69 P.J. Garrahan, J.M. Glynn: Nature *211*, 1414 (1966)
8.70 P. Mitchell: Biol. Rev. *41*, 445 (1966)
8.71 P.D. Boyer: FEBS Lett. *58*, 1 (1975)
8.72 B. Annajev, V.K. Koltover, L.M. Raichman, V.I. Suskina: Dokl. Akad. Nauk
      SSSR *196*, 969 (1971)

8.73 Yu.G. Molotkovsky, V.S. Dzubenko, V.N. Timonina: Fisiol. Rast. *19*, 525 (1972)

8.74 B. Chance, C.P. Lee, T. Ohnishi, J. Higgins: In *Electron Transport and Energy Conservation* (Adriatica Editrice, Bari 1970) p.29

8.75 B. Chance, M. Pring, A. Azzi, C.P. Lee, L. Mela: Biophys. J. *9*, A-90 (1969)

8.76 V.K. Koltover, L.M. Reichman, A.A. Jasaitis, L.A. Blumenfeld: Biochim. Biophys. Acta *234*, 306 (1971)

8.77 L. Packer: J. Biol. Chem. *235*, 242 (1960)

8.78 L. Packer: In "Book of Abstracts, VI Intern. Congr. on Photobiology", Bochum (1972) No.030

8.79 D.E. Green: Proc. Nat. Acad. Sci. USA *69*, 726 (1972)

8.80 A. Bennun: Nature New Biology *233*, 5 (1971)

8.81 L.L. Grinius, V.P. Sculachev: Biochimia *36*, 430 (1971)

8.82 V.V. Kuliene, V.P. Sculachev, A.A. Jasaitis: Biochimia *36*, 649 (1971)

8.83 E.A. Liberman, V.N. Topali, L.M. Tsofina, A.A. Jasaitis, V.P. Sculachev: Biochimia *34*, 1089 (1969)

8.84 A.A. Jasaitis: Thesis, Moscow (1972)

8.85 L.L. Grinius, A.A. Jasaitis, J.P. Kadzianskas, E.A. Liberman, V.P. Sculachev, V.P. Topali, L.M. Tsofina, M.A. Vladimirova: Biochim. Biophys. Acta *216*, 1 (1970)

8.86 L.E. Bakeeva, L.L. Grinius, A.A. Jasaitis, V.V. Kulieve, D.O. Levitsky, E.A. Liberman, I.I. Severina, V.P. Sculachev: Biochim. Biophys. Acta *234*, 177 (1971)

8.87 Y. Barber: FEBS Lett. *20*, 251 (1972)

8.88 B. Chance, J. Bunkenburg, J.A. McGray: Nature *225*, 705 (1970)

8.89 W. Junge: In "Book of Abstracts, VI Intern. Congr. on Photobiology", Bochum (1972) No.034

8.90 E.A. Liberman, V.P. Topali, L.M. Tsofina, A.A. Jasaitis, V.P. Sculachev: Nature *222*, 1076 (1969)

8.91 H. Rottenberg, T. Grunwald, M. Avron: Eur. J. Biochem. *25*, 24 (1972)

8.92 Sh. Shuldiner, H. Rottenberg, M. Avron: Eur. J. Biochem. *25*, 64 (1972)

8.93 H. Rottenberg, T. Grunwald: Eur. J. Biochem. *25*, 71 (1972)

8.94 A.T. Jagendorf, E. Uribe: Proc. JEG *1*, 678 (1966)

8.95 A.T. Jagendorf, E. Uribe: Proc. Nat. Acad. Sci. USA *55*, 170 (1966)

8.96 E.A. Liberman, E.N. Mochova, V.P. Sculachev, V.P. Topali: Biofizika *13*, 188 (1968)

8.97 E.A. Liberman, V.P. Topali: Biofizika *13*, 1025 (1968)

8.98 V.P. Sculachev: Currents in Mod. Biol. *2*, 98 (1968)

8.99 V.P. Sculachev: In *The Energy Level and Metabolic Control in Mitochondria* (Adriatica Editrice, Bari 1969) p.283

8.100 J.T. Tupper, H. Tedeshi: Science *166*, 1539 (1969)

8.101 M. Avron: In "Book of Abstract, VI Intern. Congr. on Photobiology", Bochum (1972) No.035

8.102 S.J.D. Karlish, M. Avron: Nature *216*, 1107 (1967)

8.103 S.J.D. Karlish, M. Avron: Eur. J. Biochem. *20*, 51 (1971)

8.104 R.A. Dilley, L.P. Vernon: Proc. Nat. Acad. Sci. USA *57*, 395 (1967)

8.105 R.L. Heath: Biochim. Biophys. Acta *256*, 645 (1972)

8.106 S.J. Karlish, M. Avron: Eur. J. Biochem. *9*, 291 (1969)

*8.107 V.P. Sculachev: *Energy Transformation Within Biomembranes* (Nauka, Moscow 1972)

8.108 D.G. Nicholls: Eur. J. Biochem. *50*, 305 (1974)

8.109 T. Yamomoto, Y. Tonomura: J. Biochem. *77*, 137 (1975)

8.110 W. Junge, H.T. Witt: Z. Naturforsch. *23*, 244 (1968)

8.111 W. Junge, H. Rumberg, H. Schröder: Eur. J. Biochem. *14*, 575 (1970)

8.112 R.J.P. Williams: Ann. N.Y. Ac. Sc. *227*, 98 (1974)

8.113 G. Weber: Ann. N.Y. Ac. Sc. *227*, 486 (1974)

8.114 L.A. Blumenfeld: Biofizika *21*, 946 (1976)

8.115 L.A. Blumenfeld, V.K. Koltover: Mol. Biol. *6*, 161 (1972)

8.116 L.A. Blumenfeld, D.S. Chernavskii: J. Theor. Biol. *39*, 1 (1973)

8.117 L.A. Blumenfeld, V.I. Goldanskii, M.I. Podgoretzky, D.S. Chernavskii: Zh. Strukt. Chim. *8*, 854 (1967)

8.118 L.A. Blumenfeld, M.I. Temkin: Biofizika *7*, 731 (1962)

8.119 L.P. Kayushin, E.E. Kofman, I.N. Golubev, K.M. Lvov, M.K. Pulatova: Biofizika *6*, 20 (1961)
8.120 A.D. Makarov, A.N. Malyan, V.K. Opanasenko: Biofizika *16*, 1125 (1971)
8.121 V.P. Skulachev: Nature *198*, 444 (1963)
8.122 V.P. Skulachev: Usp. biol. Chim. *6*, 180 (1964)
8.123 D.E. Green, J. Asai, R.O. Harris, J. Penniston: Arch. Biochem. Biophys. *125*, 684 (1968)
8.124 A. Azzi, B. Chance, G. Rodda, C. Lee: Proc. Nat. Acad. Sci. USA *62*, 612 (1969)
8.125 A. Azzi, B. Chance, G. Rodda, C. Lee: Biochim. Biophys. Acta *189*, 141 (1969)
8.126 J.S. Kahn: Biochem. J. *116*, 55 (1970)
8.127 N.M. Chernavskaya, D.S. Chernavskii, L.N. Grigorov: Trudy MOIP *49*, 191 (1973)
8.128 B.F. Gray: Nature *253*, 436 (1975)
8.129 B.F. Gray, I. Gonda: J. Theor. Biol. *69*, 167 (1977)
8.130 B.F. Gray, I. Gonda: J. Theor. Biol. *69*, 187 (1977)
8.131 M. Baltscheffsky: In *The Photosynthetic Bacteria*, ed. by R.K. Clayton, W.R. Sistrom (Plenum, New York 1978) p.595
8.132 R.J.P. Williams: Biochim. Biophys. Acta *505*, 1 (1978)
8.133 H. Witt: Biochim. Biophys. Acta *505*, 355 (1979)
8.134 L.A. Blumenfeld: Quart. Rev. Biophys. *11*, 251 (1978)
8.135 B. Cartling, A. Ehrenberg: Biophys. J. *23*, 451 (1978)
8.136 M. Baltscheffsky: In *Dynamics of Energy-Transducing Membranes* (Elsevier, Amsterdam 1977) p.365

Subject Index

Dynamics of Solids and Liquids by Neutron Scattering

Editors: S. W. Lovesey, T. Springer

1977. 156 figures, 15 tables. XI, 379 pages
(Topics in Current Physics, Volume 3)
ISBN 3-540-08156-9

Contents:
S. W. Lovesey: Introduction. – H. G. Smith, N. Wakabayashi: Phonons. – B. Dorner, R. Comès: Phonons and Structural Phase Transformations. – J. W. White: Dynamics of Molecular Crystals, Polymers, and Adsorbed Species. – T. Springer: Molecular Rotations, and Diffusion in Solids, in Particular Hydrogen in Metals. – R. D. Mountain: Collective Modes in Classical Monoatomic Liquids. – S. W. Lovesey, J. M. Loveluck: Magnetic Scattering.

Neutron Diffraction

Editor: H. Dachs

1978. 138 figures, 32 tables. XIII, 357 pages
(Topics in Current Physics, Volume 6)
ISBN 3-540-08710-9

Contents:
H. Dachs: Principles of Neutron Diffraction. – J. B. Hayter: Polarized Neutrons. – P. Coppens: Combining X-Ray and Neutron Diffraction: The Study of Charge Density Distributions in Solids.– W. Prandl: The Determination of Magnetic Structures. – W. Schmatz: Disordered Structures. – P.-A. Lingård: Phase Transitions and Critical Phenomena. – G. Zaccài: Application of Neutron Diffraction to Biological Problems. – P. Chieux: Liquid Structure Investigation by Neutron Scattering. –H. Rauch, D. Petrascheck: Dynamical Neutron Diffraction and Its Application.

C. P. Slichter

Principles of Magnetic Resonance

2nd corrected printing of the 2nd revised and expanded edition. 1980. 115 figures. X, 397 pages
(Springer Series in Solid-State Sciences, Volume 1)
ISBN 3-540-08476-2

Contents:
Elements of Resonance. – Basic Theory. – Magnetic Dipolar Broadening of Rigid Lattices. – Magnetic Interactions of Nuclei with Electrons. – Spin-Lattice Relaxation and Motional Narrowing of Resonance Lines. – Spin Temperature in Magnetism and in Magnetic Resonance. – Double Resonance. – Advanced Concepts in Pulsed Magnetic Resonance. – Electric Quadrupole Effects. – Electron Spin Resonance. – Summary. – Problems. – Appendices. – Selected Bibliography. – References.– Author Index. – Subject Index.

Turbulence

Editor: P. Bradshaw

2nd corrected and updated edition. 1978. 47 figures, 4 tables. XI, 339 pages
(Topics and Applied Physics, Volume 12)
ISBN 3-540-08864-4

Contents:
P. Bradshaw: Introduction. – H. H. Fernholz: External Flow. – J. P. Johnston: Internal Flows. P. Bradshaw, J. D. Woods: Geophysical Turbulence and Buoyant Flows. – W. C. Reynolds, T. Cebeci: Calculation of Turbulent Flows. – B. E. Launder: Heat and Mass Transport. – J. J. Lumley: Two-Phase and Non-Newtonian Flows.

Springer-Verlag
Berlin
Heidelberg
New York

Digital Pattern Recognition

Editor: K. S. Fu

2nd corrected and updated edition. 1980.
59 figures, 7 tables. XI, 234 pages
(Communication and Cybernetics, Volume 10)
ISBN 3-540-10207-8

Contents:
K. S. Fu: Introduction. – *T.M. Cover, T. J. Wagner:*
Topics in Statistical Pattern Recognition. –
E. Diday, J. C. Simon: Clustering Analysis. –
K. S. Fu: Syntactic (Linguistic) Pattern Recognition. – *A. Rosenfeld, J. S. Weszka:* Picture Recognition. – *J. J. Wolf:* Speech Recognition and Understanding. – *K. S. Fu, A. Rosenfeld:* Recent Developments in Digital Pattern Recognition. – Subject
Index.

Hydrodynamic Instabilities and the Transition to Turbulence

Editors: H. L. Swinney, J. P. Gollub

1981. 81 figures, approx. 8 tables.
Approx. 340 pages
(Topics in Applied Physics, Volume 45)
ISBN 3-540-10390-2

Contents:
H. L. Swinney, J. P. Gollub: Introduction. –
O. E. Lanford: Strange Attractors and Turbulence. – *D. D. Joseph:* Hydrodynamic Stability and
Bifurcation. – *J. A. Yorke, E. D. Yorke:* Chaotic
Behavior and Fluid Dynamics. – *F. H. Busse:*
Transition to Turbulence in Rayleigh-Bénard
Convection. – *R. C. DiPrima, H. L. Swinney:*
Instabilities and Transition in Flow Between Concentric Rotating Cylinders. – *S. A. Maslowe:* Shear
Flow Instabilities and Transition. – *D. J. Tritton,
P. A. Davies:* Instabilities in Geophysical Fluid
Dynamics. – *J. Guckenheimer:* Instabilities and
Chaos in Nonhydrodynamic Systems.

T. Kohonen

Content-Addressable Memories

1980. 123 figures, 35 tables. XI, 368 pages
(Springer Series in Information Sciences,
Volume 1)
ISBN 3-540-09823-2

Contents:
Associative Memory, Content Addressing, and
Associative Recall. – Content Addressing by Software. – Logic Principles of Content-Addressable
Memories. – CAM Hardware. – The CAM as a
System Part. – Content-Addressable Processors. –
References. – Subject Index.

Turbulent Reacting Flows

Editors: P. A. Libby, F. A. Williams

1980. 38 figures, 3 tables. XIII, 243 pages
(Topics in Applied Physics, Volume 44)
ISBN 3-540-10192-6

Contents:
P. A. Libby, F. A. Williams: Fundamental Aspects. –
A. M. Mellor, C. R. Ferguson: Practical Problems in
Turbulent Reacting Flows. – *R. W. Bilger:* Turbulent Flows with Nonpremixed Reactants. –
K. N. C. Bray: Turbulent Flows with Premixed
Reactants. – *E. E. O'Brien:* The Probability
Density Function (pdf) Approach to Reacting
Turbulent. – *P. A. Libby, F. A. Williams:* Perspective and Research Tropic.

Springer-Verlag
Berlin
Heidelberg
New York